GALÁPAGOS

THE ISLANDS THAT CHANGED THE WORLD

YALE UNIVERSITY PRESS NEW HAVEN AND LONDON

Paul D. Stewart

Foreword by Richard Dawkins

GALÁPAGOS

THE ISLANDS THAT CHANGED THE WORLD

CONTENTS

FOREWORD

The good science fairy flew right round the world, looking for a favoured spot to touch with her magic wand and turn it into a scientific paradise, a geological and biological Eden, the evolutionary scientists' Arcadia. You may question her motives or her existence, but of the place she lit upon there is no doubt. It lies beneath the eastern Pacific, approximately 91 degrees west and 1 degree south, some 1170 km west of the coast of Ecuador – Darwin's 'Republic of the Equator'. She blessed this spot with her wand and turned it into a volcanic hot spot (or hell-mouth as Paul Stewart describes it). Meanwhile, her collaborator, the science fairy godmother, arranged for the Nazca tectonic plate to move in an orderly fashion towards the mainland, at a stately pace of 4 cm per year. The result of these two science-friendly circumstances – plate moving over hot spot – is that the Galápagos archipelago was spun out on what Paul Stewart calls a geological conveyor belt. It is consequently an almost perfect natural laboratory of evolution – scene of an experiment planned in scientific heaven. The experimental plots (or islands as we mortals call them) are laid out, as they should be, in order of age, from the ringing black clinker of young Fernandina in the west to Española in the east – the latter soon (by geological standards) to go the way of its vanished predecessors and disappear beneath the waves.

It is consequently an almost perfect natural laboratory of evolution – scene of an experiment planned in scientific heaven.

Richard Dawkins

I was stimulated to think along these lines, not by visiting the islands themselves – although I did go twice within one year – but by reading Paul Stewart's marvellous book. I expected it to contain some stunning photographs. How could it not, when produced by an award-winning cameraman and members of the BBC's peerless Natural History Unit – what we might call the David Attenborough stable of filmmakers? What I didn't know is that Paul Stewart has a writing style to match. The publishers sent me a typescript before I saw any of the pictures. I read it in one day (which I only do with well-written books) and closed it feeling, 'Who needs photographs when the author can paint word pictures like that?' Later the photographs arrived, and I took it all back. Maybe we don't *need* the photographs but it is wonderful to have them.

To return to my fancy of the good science fairy and her wand, the hot spot needed to be carefully placed, just the right distance from the South American mainland. Too close, and the archipelago would have been over-run by South American immigrants. Faunistically, it would just be a suburb of the mainland. Too far, and it would be too depauperate to tell us anything. But with the volcanic hell-mouth placed – as it is – at just the right distance from the mainland, and with the islands themselves neatly spaced out, all evolutionary hell breaks loose. But it breaks loose with that controlled moderation that marks the well-designed experiment: just enough richness to be interesting and revealing, not enough to confuse and smother the take-home message.

Darwin himself, as this book makes clear, didn't take home quite as much as he might have. Although he described Galápagos as the origin of all his views, those views were not at the time matured enough to motivate him to label his specimens. He muddled his finches (and the *Beagle*'s crew ate the adult tortoises) and Darwin later had to rely on the finches that FitzRoy and Covington took home, in order to reconstruct the Darwinian take-home message. The story of Darwin and Galápagos is the topic of one of the chapters. Of equal human interest is the history of the archipelago's discovery and some of the macabre tales of subsequent immigrants. So are the chapters on the geology, the life on land and in the surrounding waters, and the pressing need to conserve this priceless, natural museum of evolution – and redeem our past failings for, as Stewart eloquently puts it, 'greed and natural history repeatedly proved the twinned incubus and succubus of misery'.

Below: Reptiles, large and small, dominate these remote islands. Here, a lava lizard perches on the head of a marine iguana.

To anyone contemplating a visit to Las Encantadas – the Enchanted Isles – the Gazetteer alone is worth the price of this book. Paul Stewart's *Galápagos* will be my treasured companion on my next visit, and I shall take along an extra copy to present to the boat's library. If you are not able to go in person, reading this book and savouring its pictures is, if not a substitute, a delight to be going on with.

RICHARD DAWKINS

Richard Dawkins FRS is the Charles Simonyi Professor of the Public Understanding of Science at Oxford University. His books include The Selfish Gene *and* The Ancestor's Tale.

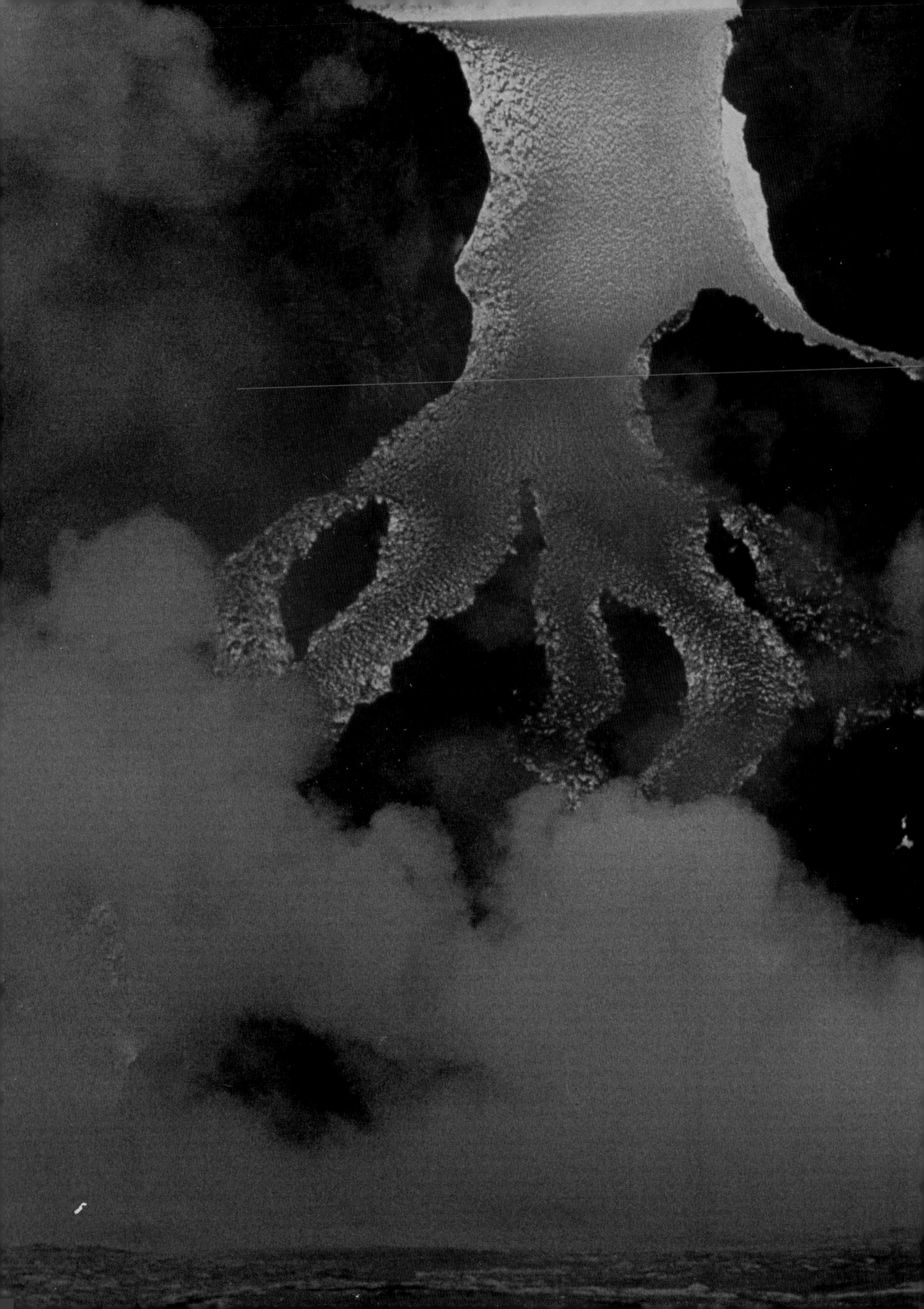

Far out in the quiet Pacific, the low-domed backs of *massive shield volcanoes* rise from the depths. Here and there they break the surface as islands. Across great spans of time *these islands have wandered*, like so many giant tortoises bathing in a pool.

The first Spanish navigators to chart the Galápagos archipelago were convinced they could see the islands move. These lands were *bewitched* – cursed even. They called them *Las Encantadas*, the Enchanted Isles: an earthly Tartarus at the furthest edge of the known world. A land fit for fiends, but not for men.

A PROLOGUE TO THE ENCHANTED ISLES

Previous page: Lava pours into the sea during an eruption on Fernandina. The existing wildlife is destroyed, but in its place come new coastlines for colonization.
Above: Darwin's arch stands at the entrance to the northernmost approach to the archipelago: a volcanic remnant carved by the elements.

THE GEOLOGICAL JOURNEY

This book is about the Galápagos's journey through time. It comprises three distinct but interwoven narratives, one geological, one biological and one human. These local histories are remarkable because they reveal important truths not just about this scattering of strange islands but also about the world beyond.

In Chapter 1 we describe the geological journey that sets each island's lifespan from volcanic birth to death and burial in the abyssal depths. Each of the Galápagos islands begins with a volcanic eruption, the meeting between fire and water that brings a new island's birth. While the Galápagos story is one of unceasing change, the great constant that gives the islands their unique identity is the Galápagos hot spot – a deep upwelling of molten magma that some believe to be rooted in the Earth's very core. 'Hot spot' is the sort of understated technical term that could only be employed to label phenomena on a planetary scale. Up close it is a hell-mouth. For many millions of years it has been lifting and puncturing the drifting oceanic plates above with periodic eruptions of lava. The hot spot has built volcanoes, islands and, ultimately, the whole Galápagos archipelago.

The island of Fernandina is the most recent progeny of the hot spot, so gives us a glimpse into the youth of the other islands. Fernandina's landscape is prone to cataclysmic change even in a human lifetime. It is harsh, unforgiving and tempestuous – and for many old Galápagos hands it is the embodied soul of the Galápagos.

To the east, in the direction of the oceanic plates' drift, we find somewhat older islands that have settled a little in their maturity. The string of volcanoes that forms Isabela's 160-km spine is still active, but subdued to the point where soil and dense vegetation have a chance to form between the conflagrations of fire and lava that periodically entomb life on the slopes.

Then we venture to the heart of the archipelago to find mature islands like Santa Cruz, which has drifted so far from the hot spot that it lies largely unfed by its magma. On high slopes clouds and rain form, and water allows thick soils and dense biological communities to thrive.

Finally, far out on the southeastern edges of the archipelago lie the oldest islands such as Española. Its volcanoes have long since been levelled by erosion and, with no elevation to condense clouds, hardly any rain falls. This is a true desert island, with communities of saddle-back 'giraffe' tortoises and cactuses adapted to prolonged drought. But these island outposts are far from barren. In their old age they witness a final flowering as dense bird and sea-lion colonies form on their shores. They are carriers for creatures whose true home is on the oceans. Ultimately, scorched by sun and pummelled by waves, they crumble into the sea. The bulging fluid magma that in their youth pushed up the crust below them has cooled, drawing them down into the ocean as a new realm for coral fish and hunting sharks. But as we shall see in later chapters, even then their story is far from over.

If as poets say, life is a dream, I am sure in a voyage these are the visions which best serve to pass away the long night.

Charles Darwin, *A Naturalist's Voyage* (1845)

It is against this shifting geological conveyor belt that the story of life on these islands has played. Galápagos's position at the confluence of several major ocean currents, coming from nearly all points of the compass, has led to its being called a 'sea of confusion' and has greatly influenced the creatures that inhabit it. This is a place where you might meet a penguin and a flamingo walking on the same beach, or find a lost Southern elephant seal cruising a coral reef. The interplay of these currents also leads to a climatic diversity that can range from incessant rain to years of murderous drought. Among the more recent arrivals to face this natural challenge are people.

THE HUMAN JOURNEY

In CHAPTER 2 we narrate the human history of Galápagos. It's a remarkable insight into the way in which our species has changed in its attitude to the natural world in the last 500 years. Today Galápagos speaks to us about our origins. But it wasn't always like that. The natural history of the sixteenth to the early nineteenth century was dominated by the idea that plants and animals were placed on Earth by God, principally for man's use and moral education. Yet in a world made for man, what purpose could be served by islands without the means of sustaining him? Long after its discovery, Galápagos was considered a confused and perplexing place, out of kilter with the 'divine plan revealed'.

For centuries, however, these islands were frequented by men who weren't interested in such theological niceties. Pirates, whalers, murderous castaways, tyrants and convicts all arrived and attempted to make a life in a place that would ultimately lead them to tragedy. It seemed that people could find no way of sustaining themselves on these islands without killing each other or the goose that laid the golden egg. It may still be true.

In CHAPTER 3 we explain how Charles Darwin – the great naturalist whose observations on Galápagos started a revolution in scientific thought – put humankind in its rightful place on a branch rather than a pinnacle of nature and so revealed the islands as a land of riches unto themselves. By the mid-nineteenth century humans had reached the most remote places on Earth. Endemic wildlife was killed or displaced by introduced species. After our arrival, Hawaii lost 50 per cent of its terrestrial birds and New Zealand all the fabulous moas, vegetarian flightless birds up to 2 m in height. By Darwin's day such islands had so many missing or wrong pieces that the biological puzzle of life was not going to be solved there. What Darwin found on Galápagos was the last oceanic archipelago whose ecosystems were still almost intact. Why, he wondered, was the life on the Cape Verde Islands different from that of Galápagos, when the superficial appearance of the two was so similar? How had the plants and birds got there? The remarkable tree cactus and daisy plants, the huge tortoises, sea-going iguanas and penguins were wonderful examples of curious arrivals, unique species and variations on themes. Galápagos was prehistoric, a lost world for the imagination to play on. It offered real potential for explaining how life

Above: Whale oil was such a valuable commodity in the nineteenth century that ships would travel to places as remote as Galápagos in pursuit of it.

Above: Tortoises often spend the night in the warm pools of Alcedo's volcanic caldera.

works – a deep question that, if answered, would, in the words of Douglas Adams, affect the way we saw life, the universe and everything.

Darwin stated that Galápagos was 'the origin of all my views'. If it were, then we owe these islands a debt for giving us a profound insight into life itself. Many evolutionary questions, even the very existence of evolution, are debated to this day, but there is no denying that variability within a species occurs, that natural selection works with these variables and that isolation is a prime factor in allowing local environments to mould the form of organisms. The Galápagos Islands have often been described as a natural laboratory of evolution. For almost 200 years since Darwin's day they have been a focus for biological research, and in marked contrast to their early history everybody would now like to visit them.

THE BIOLOGICAL JOURNEY

In Chapters 4, 5 and 6 we introduce the wild occupants of Galápagos that gave Darwin his inspiration and continue to delight visitors today. Over unknown eons before the arrival of humans, creatures have washed up on Galápagos shores. To survive the long oceanic journey, drifting or blown across a virtual

Opposite: A brown pelican waits to feed on the fish boiled by a lava flow entering the sea near Cape Hammond, Fernandina.

desert of salt water and bare lava, takes organisms of a very particular constitution. For those lucky enough to find what they needed to start a dynasty, the journey had just begun.

Isolated from the mainland members of their own species, populations changed. Generation by generation individuals with traits that gave them a competitive edge in this new land tended to produce the most young. Their descendants tended to inherit these traits and perhaps add more variation from the chance results of mutation. These traits were to be favoured (or lost) in subsequent generations – generations that underwent yet further change when spread across the different islands of the archipelago. So via the process immortalized by Darwin as evolution through natural selection, a unique assemblage of fauna and flora became, to our eyes, stranger still. Tortoises survived as giants. Iguanas took to the sea to feed on algae. Cormorants lost the power of flight. Some fish even lost their eyes. And all across the archipelago an innocence arose or was perpetuated so that creatures that had never seen humans never came to fear them. CHAPTERS 7 and 8 reveal the continuing importance and fragility of the Galápagos Islands in the modern world. We see that the urge to over-exploit them is far from being a thing of the past, and that being 'loved too much' might result in their losing the basic foundation of their uniqueness – their isolation. Galápagos today offers a different kind of natural experiment to that which Darwin observed. It tests whether we really have the resolve to make good our modern commitment to conservation and sustainable living. Despite all the challenges, for the moment at least, the Galápagos fauna and flora remain a spectacular reminder of a planet before people.

In the GAZETTEER we provide practical information on visiting the islands or helping their conservation. Use it to add your own Galápagos journey to the ones you have read, and perhaps see the islands for yourself.

This book reveals that while Galápagos may not fit our ideals of a biblical Eden, it is a wildlife paradise and a refuge for the human spirit. It is a little world within itself (as Darwin said), a working scale model, where the most fundamental processes of evolution, ecology and even human history are showcased in a way that can be understood by a less than omniscient observer. A mental voyage around Galápagos takes us through biological history from cooling land devoid of life to lush, diverse, ecosystems of complex interdependency. It also takes us through human history from the age of enlightenment, through romance to the age of reason. It charts human attitudes to wildlife from straightforward exploitation to near spiritual reverence. From a time when wild resources seemed inexhaustible to a time when our own future seems threatened by their loss.

1

The Galápagos are no ordinary islands. Everything about them seems *otherworldly*. Rugged and imposing, they are huge shield volcanoes, plumbed directly into the heart of the Earth, rising from *great oceanic depths* aboard a giant submarine platform, their summits puncturing *the vastness* of the eastern Pacific.

ISLAND ORIGINS

The Galápagos islands are the product of one of the most volcanically active regions on our planet, a 'hot spot' situated almost 1000 km west of the South American coastline, right on the equator and, as fate would have it, at the confluence of four major ocean currents. Through time, wave and wind have cast a very unusual collection of life on to their shores.

RESTLESS GIANTS

Galápagos is an archipelago of 13 main islands (greater than 10 sq km), six smaller ones and well over 100 islets and rocks, scattered like coals over 430 km of open ocean, from Darwin Island in the northwest to Española in the southeast. The total land area is about 8000 sq km, and more than half of this belongs to the island of Isabela, by far the largest of them all.

Previous page: A pair of tuff cones on Santiago, where fine grains of volcanic ash have been cemented together. Above: A NASA satellite image of the Galápagos archipelago.

Around 90 per cent of the world's volcanic activity occurs at tectonic boundaries, where oceanic plates are moving apart, like the mid-Atlantic ridge, or in places where they are carried beneath giant continental plates, such as along the western edge of South America, resulting in the spectacular Andes mountains. But the Galápagos islands are the product of a very different set of forces. Along with other oceanic islands such as Hawaii, the Azores and Réunion, they are the result of a 'hot spot' or mantle plume, where superheated malleable rock, less dense than its surroundings, rises up from deep within the Earth in a fixed location.

As the 'soft rock' plume nears the surface, at a depth of about 100 km, the pressure diminishes and it begins to melt, with temperatures reaching 1400°C. The melt separates from the rock to form magma, which punctures the brittle uppermost shell of the Earth, known as the lithosphere. Here, several kilometres beneath the Galápagos archipelago, the magma becomes trapped in large chambers. But every now and then, when the pressure gets too much, it is able to force its way to the surface, resulting in a spectacular volcanic eruption.

The demon of fire seemed rushing to the embraces of Neptune; and dreadful indeed was the uproar occasioned by their meeting. The ocean boiled and roared and bellowed, as if a civil war had broken out in the Tartarean gulf.

Benjamin Morrell,
A Narrative of Four Voyages (1832)

In the last 200 years some 60 eruptions have been recorded from eight different Galápagos volcanoes. The most recent was that of Sierra Negra on the island of Isabela in October 2005. It's the largest volcano in the archipelago, 11 km in diameter, and it erupted with fountains of lava jetting over 200 m high, and a vast volcanic cloud rising 12 km into the sky. On the first day the volcano was producing well over a million cubic metres of lava per hour and the eruption, although eventually slowing, lasted for a week. Others in recent memory have continued for months, so it's little wonder these islands grow so quickly, in geological terms at least.

The youngest and most active of all the main islands is Fernandina. It's less than 700,000 years old, and fresh evidence suggests it may have raised its head above the waves as little as 30,000 years ago. Since then it has grown 1.5 km tall, with at least 25 eruptions in the past two centuries. The caldera, measuring 6 km wide and 850 m deep, has been created by the periodic evacuation of magma from the chamber beneath. In 1968 a monumental event occurred here. A few weeks after a brief eruption on its flank, two-thirds of its caldera, including a large beautiful lake full of ducks, collapsed by about 300 m into a giant magma void, casting 1.5 cubic kilometres of gas and magma out from below. Ash was lifted more than 25 km, killing vegetation over 8 km from the rim. Overnight the whole profile of the island changed. Shock waves recorded in

Above: Land iguanas migrate to nest right on the rim of Fernandina volcano.

Colorado, USA, were deemed comparable in magnitude to the largest nuclear detonations.

Eruption after eruption builds a volcano and ultimately an island. Magma rising from the hot spot heats the lithosphere beneath, causing it to expand upwards. And this, together with the thickening of the Earth's crust, has helped to create the Galápagos platform, an unusually shallow region of the Pacific upon which the archipelago sits. Peeling back the ocean offers a whole new perspective on the dramatic topography, with the islands revealed to be simply the summits of vast undersea volcanoes. Roca Redonda, an offshore promontory 30 km to the northwest of Isabela, is only 60 m above sea level, yet this is just the tip of a huge mountain 20 km across at the base and 3 km tall!

This concept of stripping away the ocean has been partly achieved in history. Over the past 2–3 million years, during the last great ice age, sea levels would have been much lower, perhaps by as much as 120 m. Many parts of Galápagos that are submerged today would have been exposed as arid lowlands, and an ice-age map might have shown Fernandina and Isabela joined, a trail of island stepping stones between Santiago and Daphne Major, with the latter surely connected to Santa Cruz.

A VOLCANIC SPECTACLE, 1825

Benjamin Morrell visited Galápagos in 1825 as captain of the schooner *Tartar* seeking new sealing grounds. On 14 February, while anchored in Banks Bay in northern Isabela, he and his crew encountered, and barely survived, an almighty eruption on Fernandina. He left us with this evocative eye-witness account.

'Monday the fourteenth, at two o'clock, am, while the sable mantle of night was yet spread over the mighty Pacific, shrouding the neighbouring islands from our view, and while the stillness of death reigned everywhere about us, our ears were suddenly assailed by a sound that could only be equalled by ten thousand thunders bursting upon the air at once; while, at the same instant, the whole hemisphere was lighted up with a horrid glare that might have appalled the stoutest heart! I soon ascertained that one of the volcanoes of Narborough island [Fernandina], which had quietly slept for the last ten years, had suddenly broken forth with accumulated vengeance ... The heavens appeared to be one blaze of fire, intermingled with millions of falling stars and meteors; while the flames shot upward from the peak of Narborough to the height of at least two thousand feet ...

'But the most splendid and interesting scene of this spectacle was yet to be exhibited. At about half-past four o'clock am, the boiling contents of the tremendous caldron had swollen to the brim, and poured over the edge of the crater in a cataract of liquid fire. A river of melted lava was now seen rushing down the side of the mountain, pursuing a serpentine course to the sea, about three miles from the blazing orifice of the volcano. This dazzling stream descended in a gully, one-fourth of a mile in width, presenting the appearance of a tremendous torrent of melted iron running from the furnace ...

'Our situation was every hour becoming more critical and alarming. Not a breath of air was stirring to fill a sail, had we attempted to escape; so that we were compelled to remain idle and unwilling spectators of a pyrotechnic exhibition. All that day the fires continued to rage with unabating activity, while the mountain still continued to belch forth its melted entrails in an unceasing cataract.

'The mercury continued to rise till four pm, when the temperature of the air had increased to 123° and that of the water to 105°. Our respiration now became difficult, and several of the crew complained of extreme faintness. It was evident that something must be done and that promptly, 'O for a cap-full of wind!' was the prayer of each. The breath of a light zephyr from the continent, scarcely perceptible to the cheek, was at last announced as the welcome signal for the word, 'All hands, unmoor!' ... The *Tartar* slid along through the almost boiling ocean at the rate of about seven miles an hour.

'On passing the currents of melted lava, I became apprehensive that I should lose some of my men, as the influence of the heat was so great that several of them were incapable of standing ... Had the wind deserted us here, the consequences must have been horrible. But the mercy of Providence was still extended toward us – the refreshing breeze urged us forward towards a more temperate atmosphere ... We now steered for Charles island [Floreana] ... and came to anchor in its northwest harbour at eleven pm. Fifty miles and more to the leeward, in the northwest, the crater of Narborough appeared like a colossal beacon-light, shooting its vengeful flames high into the gloomy atmosphere, with a rumbling noise like distant thunder.'

A WORLD IN MOTION

The Galápagos islands are in a constant state of flux. And they are not just building through eruption and falling through erosion. They are also travelling. The hot spot where they are born lies beneath the northern edge of the giant Nazca plate, which is shifting southeast towards South America at about 4 cm a year. This may not seem fast, but in a million years – a blink of a geological eye – the islands are carried some 40 km on this geological conveyor belt. As they distance themselves from the hot spot they become increasingly inactive. They, and the crust below, cool and contract. The islands slowly sink.

In the far southeastern corner of the archipelago lies Española, at the very end of the Galápagos production line. It is believed to be the northern section of a once much larger volcano, and the lavas found here date as being 3–4 million years old, making it the oldest island of them all. Pummelled by the erosive forces of wind and wave, it is now a shadow of its former self, standing little more than 200 m tall. It gives us a sense of what all the grander, more active Galápagos islands may look like in time, as they age, sink and erode.

Millions of years from now every Galápagos island we see today will be drawn beneath the waves. Direct evidence lies to the east of the archipelago, where sunken volcanoes, called 'seamounts', can be found, deep beneath the waves. This chain of submerged ancestral islands is known as the Carnegie Ridge, and it stretches

THE DEEP HISTORY

Studies of the composition of volcanic rocks from Malpelo Island, lying on the Nazca plate and part of the Galápagos hot-spot 'trail', as well as igneous rock complexes along the Pacific margin of Costa Rica and Panama, shed fascinating light on the deep history of the Galápagos hot spot.

Some of the most unusual igneous rocks found in Central America consist of portions of ocean island, seamount volcanoes and ridges 20–71 million years old (mya). Their age and composition directly link the Galápagos hot-spot trails on the Pacific Ocean floor with the Caribbean 'large igneous province', an oceanic plateau where the hot head of a magma plume had spread out beneath the lithosphere. It's thought to be around 72–95 mya.

If that's the case, then the Galápagos hot spot may have played a fundamental role in helping to build landbridges between the Americas during the late Cretaceous–Paleocene period (*c.* 100–54 mya), and, much more recently, during the Pliocene–Holocene period (5.3 mya–present), allowing the flow of terrestrial fauna such as mastodons, sabre-tooth cats and ground sloths between North and South America, and blocking the movement of marine life between the Pacific and the Atlantic.

all the way east to the Peru–Chile trench, where the whole ocean floor is sliding beneath the South American continent, lifting the mighty Andes.

But this 'conveyor belt' is not the whole story. For, unlike Hawaii, where a steady stream of islands has been generated by a mantle plume for the past 80 million years, in Galápagos other, more complex geological forces are at work. The lavas here contain an unusual mix of chemical compositions because the Galápagos hot spot lies just 100 km or so south of an active mid-ocean ridge known as the Galápagos rift or spreading centre (see illustration, page 26), which is gradually moving apart and producing great volumes of its own magma.

The Galápagos spreading centre has been constantly shifting position ever since its formation 23 million years ago (mya), because of major plate reorganization. It's believed that between 20 and 12 mya the volcanic hot spot was situated at the spreading centre; from 12 to 7.5 mya it lay beneath the Cocos plate to the north and from then until now under the Nazca plate. This also explains why the Cocos plate has its own chain of ancient seamounts, which are travelling northeast at about 8 cm a year.

Above: Underwater gas vents (fumaroles) on the ocean floor, Roca Redonda.

Because the Cocos and Carnegie ridges are eventually carried beneath the plates along the western flanks of Central and South America, it's hard to estimate the actual age of the Galápagos hot spot. One of the ancient seamounts discovered in the early 1990s, 1500 m below sea level on the Carnegie ridge, had round cobbles on a flat top, proving that it was once an island eroded by wave action, and it's thought to be around 8 million years old. But some believe the Galápagos hot spot may be considerably older than that (see box opposite).

Previous page: A pair of courting waved albatrosses. Almost the entire world population of this beautiful bird nests on Española, in the ancient southeastern part of the archipelago.
Right: Complex tectonic plates and ridges surround Galápagos. The islands lie on the Nazca plate, just to the south of the Galápagos rift or 'spreading centre', which is slowly shunting towards the southeast.
Opposite: Lava fountains from a spatter cone on Cerro Azul, Isabela.

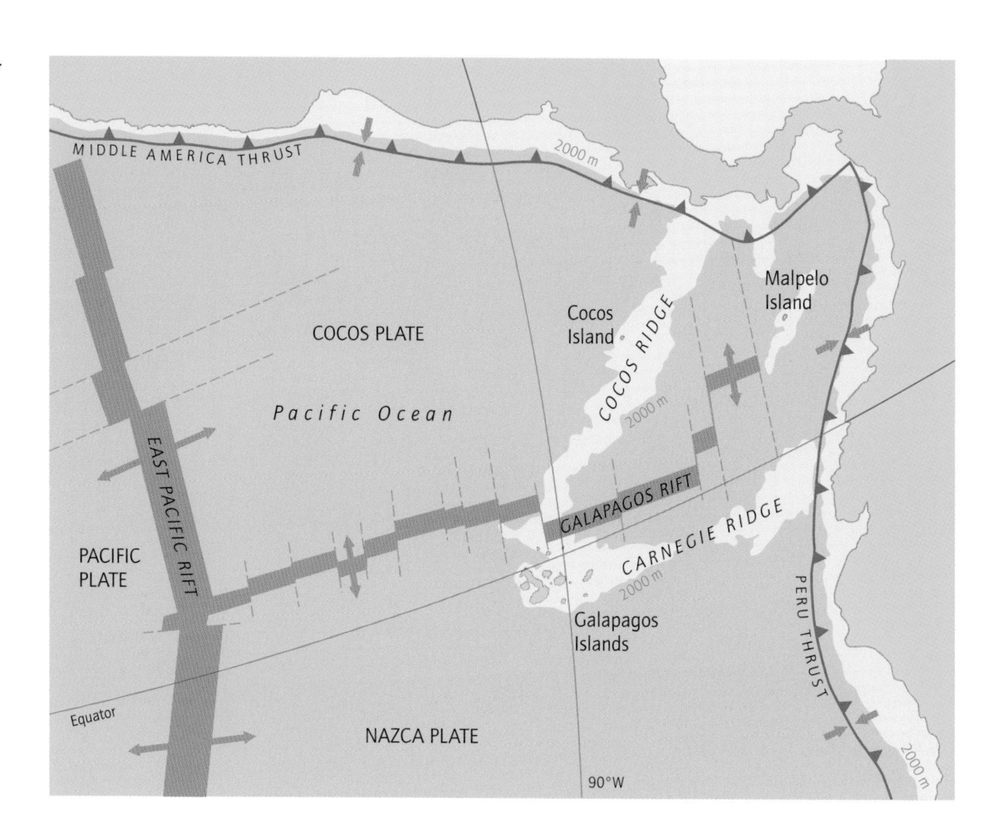

VOLCANIC LEGACY

Most Galápagos eruptions are relatively mild by world standards, with generally large lava flows rather than mighty explosions. This is because the molten rock arising from the Galápagos hot spot is mainly basaltic, with a low concentration of silica and water. But there are exceptions. On Alcedo volcano, on Isabela, the lava is anomalously rich in silica, clues of which lie in the glass-like black obsidian found here. Silica can thicken the mix, causing violent explosions. When Alcedo erupted 100,000 years ago it blew in the style of Mount St Helens, depositing huge blankets of pumice, which wiped out most of its native flora and fauna. Today this pumice can be found washed up on shores as distant as Española.

In the west of the Galápagos, the volcanoes tend to be shaped like overturned soup bowls, with rounded tops and a decreasing slope gradient. These are called shield volcanoes. The eruptive vents that feed them tend to occur either on 'circumferential' fissures around the flat summits or on 'radial' fissures – rather like the spokes of a bicycle wheel – on the lower flanks of the volcanoes.

The Galápagos archipelago is littered with clues to its turbulent past. When fluid magma emerges at temperatures of around 1100°C, the fiery lava flows,

travelling at anything up to 40 km an hour – particularly those close to the vent and on the steepest slopes – begin to cool and solidify. A crust gradually forms with two distinctive patterns. Aa (after the Hawaiian for 'stony with rough') lava is sharp and rubbly and tends to form on steeper terrain with high magma discharges. It often bulldozes piles of 'clinker' rubble ahead as is commonly seen along the coast of Fernandina. Pahoehoe (after the Hawaiian for 'smooth, unbroken') lava is formed on more gradual flows. It is thin-skinned and when it slows down the molten lava beneath buckles, giving it a more intestinal or ribbed appearance. Some of the most dramatic examples can be seen at Sullivan Bay in the east of Santiago. Indeed flowing lava creates all sorts of interesting features. Small 'hornitos' or 'driblet' cones are formed by pressurized lava squirting up through breaks in the lava's crust. And 'lava tubes' can result when the exterior of a flow solidifies and the liquid interior drains out downhill. On Santa Cruz large lava tunnels can be found, up to 10 m wide and several kilometres long.

During an eruption, two sorts of cones may be built around volcanic vents as lava, thrown into the air, is torn by expanding gases into fluid hot clots, up to 50 cm across. When the spatter hits the ground these clots weld together to form spatter cones. Cinder or scoria cones form when the fragments of lava solidify in the air after they are ejected from a single vent and build around it. These particles, known as tephra, vary in size from a few millimetres (ash particles) or a few centimetres (lapilli particles) to giant lumps called bombs. Many scoria cones are also termed parasitic or satellite cones, as they are fed from the main volcano and are usually found on fissures and faults. Nearer the coast, or offshore from a major volcano, different types of parasitic cone can be formed, known as tuff cones or tuff rings. This happens when molten rock meets water, causing highly fragmented lava to be explosively flung about in small particles, which later cement together into a harder rock called tuff.

The whole pattern of volcanoes on Galápagos is overlaid by a series of faults, or weakened splits in the upper crust. In places these faults drop dry land beneath the waves, and in others lift the seabed above the water. The cliffs on northeast Santa Cruz contain old beach deposits and corals, and the channel that separates Baltra and North Seymour from Santa Cruz owes its existence to the same fault system, as do the Plazas, two small islands off the northeast coast of Santa Cruz.

But one of the most dramatic outcomes of a likely fault event is best seen on a cruise past Volcán Ecuador, in the extreme north of Isabela. This volcano appears to have one half missing. It is possible that this is the result of gradual erosion, or even an eccentric formation, but many believe it is more likely that the entire western half

Top left: Pahoehoe lava, Santiago. Top right: A close-up of a burst bubble in pahoehoe lava – the petrified record of a fleeting moment. Bottom: Debris from cinder cones surrounded by lava fields, Santiago.

Above: An opuntia cactus forest, Santa Fe. It was arid landscapes like this that deterred human settlement for almost 300 years after the islands' discovery.

sheared off catastrophically along a fault into the ocean. Such a titanic event would have sent massive tidal waves through the region. Indeed the possibility that tidal waves periodically sweep this archipelago as a result of collapses and sea eruptions is an interesting one. They would undoubtedly result in many animals such as tortoises and iguanas being swept into the seas and perhaps to other islands.

CASTAWAYS AND COLONIZERS

Most first-time visitors to Galápagos, after leaving the lush Ecuadorian mainland around Guayaquil, are surprised by how grey, brown and black the islands appear.

Lying in the Pacific dry belt, most of the terrain here is semi-desert lowland, making it a tough challenge for colonizing plants. Bizarre-looking prickly pear and candelabra cactuses, flame and palo santo trees pepper the rugged landscape. Only on some of the larger islands, with much higher elevation, does enough rain fall to allow evergreen scalesia trees, miconia shrubs, club mosses, ferns, grasses and sedges to flourish. (We shall see more of these in Chapter 4.) Today there over 850 species, subspecies and varieties of vascular plants in the archipelago – relatively few compared to the 20,000 or so species found on mainland

Above: Salt-tolerant mangroves were early colonizers of Galápagos shores, carried here by ocean currents.

Ecuador, but they make for an odd assortment. Those with large flowers or heavy seeds, like conifers, are distinctly absent in Galápagos. The same is true of plants that have co-evolved with a specialist pollinator, or perhaps have a symbiotic relationship with fungus, both of which apply to orchids – only 14 different orchids are found in Galápagos, although there are thousands of species in mainland Ecuador.

The key reason for this is the vast oceanic barrier that separates the Galápagos archipelago from the South American mainland, as it has done since the ancestors of these islands first raised their heads above the water tens of millions of years ago. For a plant, getting to the islands involves serious long-distance dispersal with four possible means of transport: on the wind, on the waves, by birds and, more recently, by human introduction, both accidental and intentional (see Chapter 7).

Both the northeast and southeast trade winds blow toward Galápagos from the American mainland, and over the millennia they have carried the lightweight spores of ferns, mosses and lichens, as well as algae, bacteria and fungi, to the islands, with rain showers helping with the drop-off. Though most vascular plants have heavier seeds, some, like members of *Asteraceae*, the sunflower family, have fruits with plume-like hairs that allow them to be picked up by the wind. Dozens of these species have found their way to Galápagos. Orchids, too, with seeds the size of dust particles, would have been blown in on the wind.

Ocean currents, like the cold Humboldt current flowing northwest up the flanks of South America and the warmer Panama Flow sweeping in from Central America, have brought ashore their own cargo of floating, salt-tolerant seeds and fruits, such as those of the sea grape or inkberry, the saltwort, the beach morning glory and the black mangrove.

Of the 560 or more indigenous plant species in Galápagos, about 180 are endemic – found nowhere else in the world. It's thought that birds are responsible for introducing more than half of all indigenous plants to the islands, mainly by carrying seeds and fruits in their stomachs but also by having them stuck to their feathers or glued to mud on their feet. Even if a seed does manage to find its way here, there are all sorts of other hurdles to overcome, especially if it's dropped at the saline littoral zone or in the arid zone just inland. With so little soil and rainfall a plant faces a real battle to establish itself. Most will simply perish. Dormancy is a definite advantage, and it's likely that many of the successful colonizing plants were able to wait until conditions became more attractive for germination, such as during an intense El Niño period, when rainfall dramatically increases. (We shall hear more about El Niño later in the chapter.)

Sea and air have also brought a very unusual mix of animal life to the shores of Galápagos. Strong swimmers, like sea lions, fur seals, penguins and green turtles, and good fliers, such as albatrosses, shearwaters and tropicbirds, have made their own way here, helped by the currents and prevailing winds, from a range of different locations. Galápagos sea lions have their origins in the northern hemisphere and are related to the playful Californian sea lion. Greater flamingos probably came from the Caribbean. Galápagos fur seals are closely related to the southern fur seals, which would have made their way here from southern South America. And Galápagos penguins are related to Humboldt penguins, which live along the coast of Chile – and from which they probably evolved – as well as Magellanic penguins, which live off the east and west coasts of South America.

Though coastal Galápagos hums with both migrant and resident life, on land there is a paucity of mammals, birds and invertebrates. Today there are just two species of bat, four species of rice rat and 29 resident species of land birds, of which 22 are endemic to the islands and include the Galápagos hawks, rails, doves, mockingbirds and finches. And although there are over 1700 native insect

THE ORIGINS OF THE GALÁPAGOS GIANT TORTOISES

Left: A giant tortoise on the rim of Alcedo volcano, Isabela.

Giant tortoises, on the back of which the archipelago earned its famous name (see CHAPTER 2), are among the largest of their kind in the world, weighing up to 250 kg. So how did the ancestors of these iconic animals get here and from where did they come? This has long been debated, as have the systematic relationships of these tortoises, ever since Darwin's visit over 150 years ago. Most believe the first tortoise colonists arrived in Galápagos by rafting in on vegetation from South America, already with a helpful predisposition to gigantism. Evidence to support the latter part of this theory has been found in the Seychelles, where giant tortoises have colonized at least three times. Giant tortoise fossils are also known from mainland South America.

By studying the mitochondrial DNA of Galápagos giant tortoises, *Geochelone nigra*, and related subspecies from mainland South America, scientists have discovered that the closest living relative is *Geochelone chilensis*, or Chaco tortoise, which lives in dry lowland habitats in Bolivia, Paraguay, Argentina and Patagonia. Evidence from time estimates, based on a molecular clock (a technique used in genetics to date when two species diverge, deducing elapsed time from the number of minor differences in DNA sequences), suggests that the split between *G. chilensis* and the Galápagos lineage probably happened 6–12 million years ago, before the oldest existing Galápagos island came into being, and it is thought to have occurred on mainland South America. And although no one really knows when giant tortoises first arrived in Galápagos, the oldest subspecies split within *G. nigra* is estimated to have happened no more than 2 million years ago, consistent with diversification on the existing islands.

species, this is still very few compared to the mainland.

Swept here on freak storms or carried in on rafts of vegetation during flash floods, the ancestors of all these modern-day residents would have faced a truly perilous journey to reach these remote islands, with dehydration and exhaustion taking their toll. Perhaps it should come as no surprise, then, that the animals that have most successfully made this crossing and carved a comfortable niche for themselves in Galápagos are reptiles. Scale-covered and cold-blooded, reptiles are much more energy efficient than mammals; they can endure extreme temperature change, are salt tolerant and can survive long periods without food and water. Some 22 species from five different families survive here today and these include marine turtles, iguanas, geckos, snakes and the famous giant tortoises.

As with the giant tortoises (see box, page 33), studies of molecular data suggest that the species divergence of Darwin's finches – of whom we shall hear more in CHAPTER 3 – may have occurred during the lifespan of the existing islands, but that endemic marine and land iguanas could have split from each other as much as 10–20 million years ago. It's been suggested this happened within the archipelago, on islands that have long since submerged.

A FICKLE CLIMATE

The Galápagos archipelago is so far removed from any continental landmass that its climate is largely determined by the complex pattern of ocean currents that sweep its shores, driven by the trade winds. Conditions are never predictable and are often severe.

There are two broad seasons in Galápagos. June to November is known as the garúa season, when the air is generally cooler and foggier and the seas more choppy, under the influence of the cool Humboldt current and the southeast trade winds. Temperatures can fall to 20°C. The air cooled by the ocean currents literally forces the warm air upwards, creating an 'inversion layer' and cloaking the highlands in drizzley mist or *garúa*. Confusingly, this is also often known as the dry season, because the coastal lowlands receive precious little rainfall – maybe 10 mm a month. Between December and May is the warm season, known to sailors as the doldrums, when the southeast winds slacken, warmer waters from the Panama Basin move in, the skies clear and temperatures can rise above 32°C. There's much more of a tropical feel to the islands, interrupted by the occasional heavy downpour. It's at this time of year that most plant growth happens and land animals breed.

But about every three to seven years, for a period of anything from six to 18 months, this climatic pattern is dramatically disrupted. There is no 'sea of confusion'

Above: Fur seals playing in a pool at Cape Hammond, Fernandina.

about temperature zones when El Niño swells into Galápagos waters. The sea becomes uniformly tropical, which, for most creatures reliant on phytoplankton-rich oceans, is a catastrophe of dire proportions. The cool Cromwell and Humboldt currents that nourish life vanish, replaced by warm, clear waters that are almost entirely devoid of phytoplankton. Starvation hits whole populations, often whole species, with astonishing mortalities.

In the past 40 years there have been five big El Niño events in Galápagos, the most recent in 1997–8. Air temperatures were 4–5°C above average, 330 cm of rain was recorded over the two-year period at the Charles Darwin Research Station (as opposed to the usual 20–40 cm) and terrestrial plants and animals flourished. Darwin's finches on Daphne Major raised clutch after clutch, with anything up to five eggs at a time, which is exceedingly rare in a normal year.

But for the marine life it is very different. Animals like fur seals adapted from colder climes and living on a delicate margin of ecological tolerance lost a third of their numbers in 1982–3, while in 1997–8 sea-lion populations in the central and southern colonies declined by 48 per cent due to starvation and migration. Many marine iguanas perished for want of suitable algae on which to graze and seabirds such as blue-footed boobies failed to rear their young, because of flooding and starvation.

These chaotic reversals in fortune between land and sea are a natural feature of Galápagos and historically there are ample 'typical' years for hard-hit species to recover their numbers. However, recent global climate changes may be making the pendulum swing too rapidly for some rarer species to cope. And although all the greenery may look picturesque, the constant rain is quite unpleasant for people: it is reported that in 1982–3 there was a sharp increase in separations and divorces on Galápagos.

El Niño means 'boy child' or 'Christ Child', a name given to the event by Spanish-speaking fishermen because it tends to begin around Christmas. The flip side to El Niño, which often follows it, is La Niña ('girl child'), when quite the reverse happens: both air and water become cooler than normal. This is really beneficial to marine life, but on land the drought-like conditions can take a serious toll on animals and plants.

The ancient El Niño–La Niña cycle is all part of the challenging rhythm of life in Galápagos, and most plants and animals find ways to cope with, and adapt to, the very unpredictable climate out here. But some specialists, like the rare Galápagos flightless cormorants, are particularly vulnerable in El Niño years. Fewer than a thousand of these birds inhabit small stretches of coastline on Isabela and Fernandina, where they feed on the ocean floor, searching under rocks and in crevices for octopuses and bottom-dwelling fish such as eels. Because of this specialist diet and because they have a high metabolism and cannot fast for more than two to three days, any food shortages can cause serious problems. In 1983 cormorant numbers crashed from 850 to 400, all chicks and juveniles were abandoned and only a small core population of adults survived, although they did manage to rebound to their original number in less than a year.

The same fickle currents that, in some years, feed this remote archipelago with torrential downpours and in others starve them through desert drought have also, over the millennia, brought a strange collection of life to their shores. And they've had one other profound impact.

Opposite: In the sea mists of coastal Fernandina, two flightless cormorants court, one presenting the other with a gift of seaweed.

For in February 1535, while sailing from Panama to Peru, the then bishop of Panama, Fray Tomás de Berlanga, found his ship becalmed off the coast of South America. There was little he could do. He and his crew drifted into the Pacific, carried by the Panama current into waters unknown.

Little did the bishop know it, but two weeks later, on 10 March, he would be the first person in recorded history to set foot on the Galápagos islands.

2

The Chinese say, *'Happy is the land with no history.'* Few islands can have had as little history as the Galápagos, but it has crammed in its share of *misery* all the same. When the first seafarers concluded that man was not supposed to be here, they had a point. Every human venture begun on the Galápagos, virtually without exception, has ended in *failure and tragedy.* Any attempt to exploit the resources of this land soon runs into trouble. Then from the very Darwinian instincts of self-interest and self-preservation come *Machiavellian thoughts* and *Macbethian deeds.* To trace the origins of these islands' human curse, and their ultimate redemption, we must go back to their discovery.

HUMAN DISCOVERY

WHERE 'GOD HAD SHOWERED STONES'

Previous page: Small volcanic islands like Sombrero Chino near Santiago were a far cry from the 'paradise islands' early travellers hoped to find at these latitudes. Above: Seen from a distance, the islands' beauty belies their inhospitable nature.

Some might say that it was ironic that Galápagos, the inspiration for an idea that at one time was seen to threaten the very foundation of Christian religion, was itself discovered by the church.

Fray Tomás de Berlanga, bishop of Panama, had been instructed by the Holy Roman Emperor Charles V, king of Spain, to journey to Peru to unravel the feuding of the conquistadors over their new empire. He was also to report to the emperor on the nature of the new possessions, and to try to moderate the barbaric treatment of the indigenous people at the hand of these ruthless soldiers of fortune.

Tomás raised a crew to man a light caravel, a fast design of sailing ship preferred by the Spanish and Portuguese for long-range voyages. He included all the food and provisions needed for a journey of an estimated 15 days; but of course it wasn't to be enough. Leaving harbour on 23 February 1535 he sailed under a good wind for seven days towards Peru. To navigate, his ship ran close to the coast, but on the eighth day it was becalmed in the equatorial doldrums. As can happen at that time of year, the sun shone and neither rain nor wind came to the aid of the ship. The caravel was carried off the edge of the map, into the unknown deep Pacific.

Then on 10 March 1535 blinking eyes beheld an unknown land far ahead of them.

Tomás wrote a fascinating account of his discovery to his emperor, mentioning 'such big tortoises that each could carry a man on top of itself', 'many iguanas that are like serpents' and 'many birds like those of Spain, but so silly that they do not know how to flee'. He also described the difficulty of finding water – a recurring theme of human attempts to settle on Galápagos:

> *The boat once anchored, we all went on land and some were given charge of making a well, and others of looking for water over the island: from the well there came out water saltier than that of the sea; on land they were not able to find even a drop of water for two days, and with the thirst the people felt, they resorted to a leaf of some thistles like prickly pears, and because they were somewhat juicy, although not very tasty, we began to eat of them, and squeeze them to draw all the water from them, and drawn, it looked like slops of lye, and they drank it as if it were rose water.*

But the special curse, as one may call it, of the Encantadas, that which exalts them in desolation ... is that to them change never comes – neither the change of seasons or of sorrows.

Herman Melville, *The Encantadas* (1854)

And he noted the strange nature of the shore:

> *There were some small stones that we stepped on as we landed, and they were diamond-like stones, and others amber colored; but on the whole island I do not think that there is a place where one might sow a bushel of corn, because most of it is full of very big stones, so much so that it seems as though at some time God had showered stones; and the earth that there is, is like slag, worthless, because it does not have the virtue to create a little grass, but only some thistles, the leaf of which I said we picked.*

Many parts of Tomás's account will be familiar to a modern-day traveller to the islands. The stones like diamond and amber can still be found on some beaches – small water-worn crystals of feldspar and olivine brought up from the magmas of the deep earth during volcanic eruption. And the sour pads of prickly pear still offer water to the desperate. But these semi-precious stones, these 'thistles', these islands,

Above: Sailing ships once commonly re-supplied in Galápagos waters. Here nineteenth-century whalers meet off the 'dark jaws of Albemarle' (now Isabela).

were of little worth. The account bears none of the bravado-glow of proud discovery – indeed it was a remarkably honest piece of reporting from a time when fancy seemed to rule observation in exploration.

Most pointedly, the bishop failed to name or claim his discovery. This may have been no accident. He lived in a time when the natural world was seen to embody God's divine plan. Fundamental in that world view was that God's creation was placed here for the benefit of man. Yet on these near waterless islands virtually devoid of soil or normal sustenance there was no sign of this divine intent. What curse had caused God to 'shower stones' upon it? To make wells yield water 'saltier than the sea'? God's hand here seemed to have been restricted, once Catholic mass was said, to giving them just the water they needed to be able to leave the islands. Perhaps, to the bishop, the place had more than a whiff of sulphur about it.

Tomás's account was put away in Europe without comment until its rediscovery in a series of printed manuscripts in the nineteenth century. But more was apparently said at the time. There is a mysterious and very early vellum map, which may even bear the bishop's hand, placing and naming the islands of *Galápagos. Galápagos* is the name for a Spanish saddle, which echoed the shape of the tortoises' large dark

Above: The frigate birds' predatory maritime habits echo those of the pirates of the seventeenth century.

shells. At least two subsequent cartographers, the Flemish Gerard Mercator and Abraham Ortelius, also somehow heard of the discovery and, ever watchful for *terra nova* of the indisputably 'here be dragons' ilk, drew these islands on their influential maps in 1569 and 1570 respectively. For some unknown reason each decided there was not one, but two, Galápagos archipelagos out there in the deep Pacific.

Not that putting the islands on the map, even twice, served as any great draw to the Spanish sailors of the time. Indeed they had already given them a nickname – 'Las Encantadas'. But here was no enchantment, rather a bewitching and a curse. The curse lay in the islands' very nature. Strong currents and variable winds swept around them, over dangerously unpredictable reefs that still claim lives today. Most daunting of all, however, was the lack of predictable fresh water among the shattered lava terrain.

So time passed from Tomás's discovery – nearly 150 years – until the next main chapter in the islands' human history. In that time a myth grew up among the Spanish that the archipelago was not real at all, but rather a mirage, always elusive on a horizon of shifting mists and complex currents. In any case they had a new world of riches at their command, so what did they need with an ephemeral pile of ashes? So the Spaniards by and large kept away, a decision that was to place the islands in the sights of a group of men for whom suffering and the curse of God were little more than an occupational hazard. Enter the English West Country pirates.

'THEY WERE BUT SHADOWES'

When Francis Drake became the first privateer to enter the Pacific he declared, 'Beyond this line there be no peace.' The English had a long history of conflict with the Spanish in the New World, mostly on the Spanish Main (that part of the Caribbean through which treasure-laden Spanish galleons frequently sailed and were often ambushed by English ships). Those with papers from the English monarch to legitimize this theft were privateers, those without papers were locally known from the Caribbean as buccaneers and those without any pretence of allegiance to the English Crown or anyone else were simply pirates. Needless to say, to the Spanish they were all pirates.

In 1683 a 'merry band of men' under the captaincy of John Cook entered the Pacific round Cape Horn in a Danish ship captured and renamed the *Bachelor's Delight* (a name that loses its charming whimsy if you believe the popular assertion that among its cargo was a group of female African slaves). Among the crew was William Dampier, described as the 'mildest mannered man that ever scuttled a ship or cut a throat' or more charitably by the poet Coleridge as 'a pirate of exquisite mind'. They had no plans for Galápagos, but fate would draw them there all the same.

Below: William Dampier, 'a pirate of exquisite mind'.

Their escapade began well enough with the capture of three Spanish ships, but success bittered when they learned that a grand treasure had only just been downloaded after news of their arrival in the Pacific. The pirates had to content themselves with a cargo including a stately mule, a very large image of the Virgin and eight tonnes of quince marmalade. Hardly a treasure to bury beneath a palm tree. They also had the inconvenience of over a hundred Spanish prisoners and very little fresh water. So they sailed on to see if they could reprovision in the little islands marked on their map as Galipogas, 'which made the Spanish laugh, saying that these were inchanted islands and that there was never any but one Captain Porialto that had ever seene them …

they were but shadowes and noe reall islands.'

This may have been a ploy by the Spanish to get themselves released on the mainland, but the pirates knew the reception they could expect there. Galápagos was sought and sighted.

It is from this early pirate landing that the first detailed maps and accounts of the islands' natural history came. Men like Dampier kept journals of their observations and sealed them from the sea and salt in bamboo holders. Others like English buccaneer Ambrose Cowley made charts whose detail and care stand at odds with their authors' primary activities of sacking ports and sending other sailors to an early grave. Cowley's charts and naming of the islands long outlived him, though his natural-history account shows an understandable bias towards the archipelago's ability to cater for a hungry, thirsty crew.

Above: Early Galápagos visitors saw the islands as a wildlife buffet. William Dampier observed in his journals that even the tongue of a flamingo was a delicacy 'fit for a prince's table'.

> *Here being great plenty of provisions, as fish, sea and land tortoises, some of which weighed at least 200 pound weight, which are excellent good food. Here are also an abundance of fowls, viz. Flamingos and turtle doves; the latter wherof were so tame that they would alight upon our hats and arms, so that we could take them alive, they not fearing man, until such time as some of our company did fire at them, whereby they were rendered more shy.*

He goes on, 'But we could find no good water on any of these places, save on the Duke of York's island [now San Cristóbal].'

Little physical evidence remains of that first voyage, but today in many houses in the bustling Galápagos town of Puerto Ayora massive terracotta pots of crude earthenware are decoratively scattered in walls and alcoves. Brought from the bottom of the sea near James Bay, off Santiago, these may well be the remains of those many tonnes of quince marmalade.

PIECES OF EIGHT

William Dampier, the aforementioned 'pirate of exquisite mind', visited the islands again on the pirate ships *Duke* and *Duches*s after an uncharacteristically successful raid on Guayaquil (on the coast of modern Ecuador) in 1709. Plunder was taken to the islands to divide among the men, many of whom were in deteriorating health after contracting a plague-like disease in Guayaquil. That city's belated revenge cut the landing raiders down virtually to a man and the few survivors rapidly left 'these unfortunate islands'. Legends of hidden hoards from this and other pirate visits still fascinate many on the islands today – and it is popularly asserted that some landowners have left Galápagos with undeclared fortunes far beyond anything that farming or trade could provide. Some reliable authors go so far as to describe the garden in Puerto Ayora or the exact hole near a palm tree near Puerto Villamil in which treasure has been found. There seems little doubt that in some hidden lava tunnel or cave, out there in the Galápagos wilderness, a golden reward awaits a finder.

Above: Silver reales cob coins are the original 'pieces of eight'. They have been found on Galápagos beaches where seventeenth-century pirates dropped them.

Dampier survived not only the Guayaquil raid but many more in a lifetime that would see him become the first great travel writer and a fashionable friend of the Royal Society. He seems to have been motivated as much by his thirst to observe and record as he was by lust for Spanish bullion. We owe him a debt for some of the best early accounts of the islands, which he described as:

> *producing neither tree, herb nor grass; but a few dildo trees, except by the seaside. The dildo tree is a green prickly shrub that grows about ten or twelve feet high, without either leaf or fruit ... the Spaniards when they first discovered these islands found multitudes of guanoes and land turtle or tortoise ... I do believe there is no place in the world that*

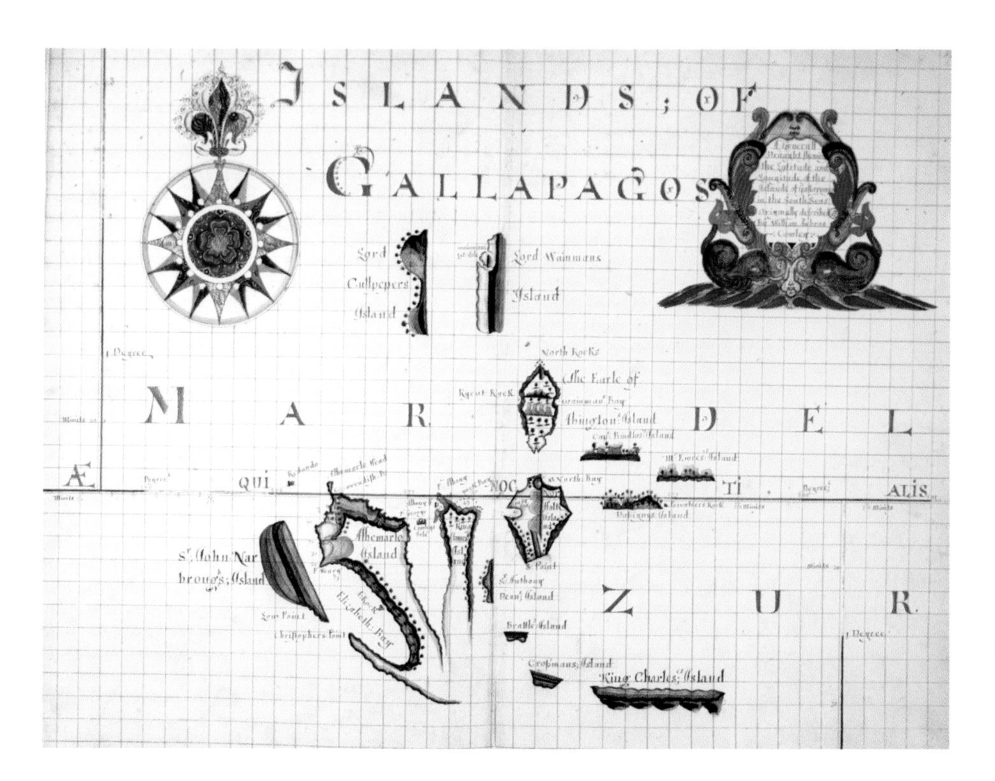

Left: The piratical Captain William Ambrose Cowley drew this surprisingly accurate early map of the islands following his pioneering voyage in 1683.

> *is so plentifully stored with these animals ... they are so tame that a man may knock down twenty in an hours time with a club.*

By the mid- to late eighteenth century the age of the pirates had drawn to a close. The next era of human involvement on Galápagos was to have a far more severe impact on the islands' wildlife. One from which it would never fully recover.

THE WHALING YEARS

Now 250 years since their discovery, the islands still had no settlers and few visitors. That was all to change in the 1790s when Enderby and Sons, a British-based whaling company, got wind of a report by Captain James Cook that there were still plenty of whales in the Pacific Ocean. By this time the Atlantic was showing steep declines in its whaling returns – the great beasts were locally going extinct. The industrial mills and cities of Europe and America were thirsty for oil, and new reserves had to be found to keep cogs turning and lights burning. So Enderby and Sons kitted up Captain James Colnett for a Royal Navy mission to the Southern Ocean to see what could be found. In his bag was a copy of Dampier's book *A New Voyage Round the World* and his Galápagos accounts. Colnett and his report were to change everything:

I frequently observed the whales leave these [Galápagos] islands and go to the westward and in a few days return with augmented numbers. I have also seen the whales coming, as it were, from the main, and passing along from the dawn of day to the night in one extended line as if they were in haste to reach the Galipagoes.

Then, as if that would not be sufficient a death knell to the islands, he added, 'This place is, in every respect, calculated for refreshment or relief for crews after a long and tedious voyage.'

Below: 'Turpining' was the name given by nineteenth-century whalers to the practice of collecting giant tortoises. Hundreds of thousands of these ancient creatures were taken from Galápagos for food.

As a result of Colnett's report, for nearly 70 years there was barely a time when commercial whalers were not somewhere in Galápagos waters. Not only did they find whales, occasional fresh water (when they knew where to look) and safe harbour but they also quickly realized the merits of tortoise as the original packaged food. The remarkable, indeed appalling, fact was that not only did tortoises taste very good but they would also survive alive and fresh in the holds of ships for several months (up to two years was reported, though hard to believe). Long after the whales had been locally cleared or deterred from Galápagos, whalers continued to come for the tortoises until an estimated 200,000 or more adults had been removed. At first they could be collected in the lowlands and on some of the smaller islands, but as these specimens and populations disappeared the men were forced to look deeper and deeper into the interior. The US whalers, who soon followed the British vessels, called it turpining (from terrapin). Soon abundant populations survived only on such precipitous volcanoes as Wolf and Alcedo, and even there the females' migration to the lowlands to lay eggs (and their lesser weight) resulted in these key members of the population being taken first.

Nor was tortoise hunting the whalers' only activity: iguanas and birds were clubbed for sport, and the Galápagos fur seals for profit. Without the islands' hidden lava grottoes, these seals would likely be extinct today. As Darwin wrote, 'What havoc the introduction of any new beast of prey must cause in a country, before the instincts of the indigenous inhabitants have become adapted to the stranger's craft or power.'

So the sailors had their fresh meat, now all they needed were a few vegetables and they had a full Sunday meal – quite different from the usual fare of ship's biscuit

and corned beef. And the need for those few vegetables helped support the islands' first settler – a man called Patrick Watkins, who embodied the human curse of these islands as few others have ever quite managed (see box, page 50).

Above: Huge pods of sperm whales, like these near Fernandina, once brought whaling fleets to Galápagos waters.

MAN OF WAR

The story of Patrick Watkins was first related by Captain Porter of the USS *Essex*, the man who virtually single-handedly brought war between Britain and the United States to Galápagos waters. It began as a dispute over the eternally contentious commodity of oil – at that time whale oil. But given the vigour with which Porter pursued his quarry one cannot help wondering if, with the American War of Independence still a recent and bitter memory, he had a few grudges to exorcize with the old colonial demon. His aim was to destroy the British whaling fleet in the Pacific. He set about this by reading British letters left in a barrel for delivery back home by returning ships (its descendant survives to this day in Post Office Bay on Floreana). From those letters he got a pretty good idea of which British whalers were where. He then left a few rendezvous invitations, which resulted in the hapless ships being seized.

A DEVIL ON THE ISLANDS

Patrick Watkins was an Irishman. Left by his ship for unruly behaviour, he set up home on Floreana near a known fresh-water spring and started farming there. He traded vegetables to passing ships – but that is only half the story. Watkins preferred his payment in grog and would spend days outside his hut in drunken stupors. His clothes were torn, his skin was burnt and his equally red hair tangled in a mass that alarmed all who met him. And well might they be afraid. Because Patrick also had ideas – to keep himself in spirits, but reduce his workload, he would waylay passing sailors until their boats had assumed them lost.

Then with force of arms, or perhaps just personality, he would engage them to do his bidding. On one attempt to do this he was caught out – a 'good whipping' on not one, but two boats was his punishment – but Patrick did not reform. Indeed he plotted a simmering revenge that would eventually get him off the islands. He invited a crew of men from the whalers *Argo* and *Cyrus* ashore for reprovision, stole one of their boats, wrecked the rest and disappeared into the island with the group of press-ganged men (who were by now probably also looking for a means of escape).

The commanders of the *Argo* and *Cyrus*, worried about Patrick's guns and a further trick, sailed on. Patrick used the stolen boat to leave Galápagos with his men, but arrived on the mainland quite alone. His island-tempered constitution may have made him a sole survivor, though years of grog make one wonder. It was generally supposed he killed his crew to save water. Some like to believe he also ate them.

Porter's tactics were amazingly successful. In just a few months he had captured a dozen British vessels with some $2.5 million of cargo, and deterred British whalers from these waters for ever.

But Porter wasn't just a military hawk – he also had a keen eye and an apparent fascination for natural history. He was the first to record that it was predominantly the female tortoises that were being caught in the lowlands during their reproductive migration and he also reported many details such as:

> *doves peculiar to these islands, of a small size and beautiful plumage were very numerous and afforded great amusement to the younger part of the crew in killing them with sticks and stones, which was nowise difficulty, as they were very tame. The English mockingbird is found in great numbers, and a small black bird with a remarkably short and strong bill and a shrill note.*

The small black birds were later immortalized as Darwin's finches, of whom we shall hear more in Chapter 3.

Above: *A post-barrel and ship's-plank 'calling cards' can still be found at Post Office Bay on Floreana, continuing a tradition begun there 200 years ago.*

Porter may also unwittingly have helped to spell the near doom of the tortoises he so enjoyed. Putting goats ashore to graze on James (now called Santiago), he discovered that they had overnight opted for freedom (as many more have since). He wrote:

> *It is probable their increase will be very rapid; and perhaps nature, whose ways are mysterious, has embraced this first opportunity of inhabiting the islands with a race of animals, who are, from their nature, almost as well enabled to withstand the want of water as the tortoises with which it now abounds.*

How right he was – by the year 2000 goats outnumbered tortoises on Santiago by over 100 to 1 and were clearly set to replace them entirely. It has cost several million dollars, but in 2006 the goats are gone and the tortoises have a future again.

In the end the first great oil war proved to have realized little. The whales were disappearing across the vast Pacific and another great source of oil to replace them had been discovered oozing from the ground in Pennsylvanian wells.

CALL ME ISHMAEL

But before the American whalers left Galápagos waters for the last time they gave passage to another notable visitor to the islands: a young whaler hand who would go on to be one of the nineteenth-century's great writers, Herman Melville, the author of *Moby Dick*.

Melville characterized the old romantic view of the islands whenever he wrote of them (and he wrote of them often enough to fill a book, *The Encantadas*, with his essays). For him they held a dark fascination that verged on the obsessive – in a way, the Galápagos were his own great white whale:

> *It is to be doubted whether any spot on earth can in desolateness furnish a parallel to this group ... while already reduced to the lees of fire, ruin itself can work little more upon them ... The Encantadas refuse to harbour even the outcasts of beasts. Man and wolf alike disown them ... No voice, no low, no howl is heard; the chief sound of life here is a hiss.*

On one point Melville was at least partly wrong. He asserts that 'the special curse, as one may call it, of the Encantadas ... is that to them change never comes; neither change of season, nor of sorrows.' But change had come to Galápagos, and its sorrows were increased. When the whalers arrived the islands were nearly pristine, but by the time they left an ecological catastrophe was under way: goats and rats were destroying places even the whalers had never reached. And 50 years later the tortoise

Opposite: Introduced goats multiply rapidly and overgraze the fragile Galápagos flora, leading to famine for goats and tortoises alike. Left: Whales are still frequently seen in Galápagos waters but their numbers may never recover to the levels of pre-whaling days.

massacre continued. They were killed as they gathered to drink at the pools of their volcanic fortresses. They were boiled down for oil to light the lamps of Guayaquil. Everyone involved knew the tortoises would disappear quickly. But there were a few good years before that time, and people have to earn a living.

REST IN PEACE

Some 300 years since their discovery and countless men lay at rest on Galápagos, the victims of plague, scurvy, duels, thirst and murder – yet not one had been born there. In bio-speak we would say that humans were transients rather than natives. That was to change near the end of the whaling era. By this time Galápagos had already started to become more of an interest for the nation of Ecuador, in whose waters it squarely lay. Under General José Villamil, a concerted and ambitious effort was made to colonize them permanently. To catalogue the many failures in this endeavour over subsequent generations is beyond the scope of this account and frankly depressing. Three main episodes will suffice to reveal how greed and natural history repeatedly proved the twinned incubus and succubus of misery.

Villamil started the first Galápagos colony on Floreana in 1832 and named it 'Haven of Peace'. The wide gulf that always seems to open between settlement names and their ultimate human outcome will become apparent as this history is told. Villamil aimed to exploit the orchilla moss *Roccella babingtonii*, used to manufacture dyes. To found the colony he had a few artisans, banished political dissenters and a band of soldiers reprieved from a death sentence for mutiny. They were later joined by the convicts of an early penal settlement and by deported prostitutes from Guayaquil. The early population would have been a few hundred.

To have pinned his hopes on these founding fathers (and mothers), it appears that Villamil was either very optimistic or not very influential. Perhaps not surprisingly he often left the place in the charge of his deputy, an Englishman by the name of Nicholas Lawson, who was there when the island's most famous visitor, Charles Darwin, arrived in 1835. The monumental impact of Darwin's brief sojourn in Galápagos is the subject of CHAPTER 3.

After five years Villamil had acquired a pack of dogs that went everywhere with him as a loyal bodyguard. The Haven of Peace became known locally as 'the Dog Kingdom'. Domestic animals went feral on the islands and pigs joined the rats and goats probably already there. And when the relatively kindly Villamil was replaced by the relatively unkindly English Colonel J. Williams, things got much worse. The island's tortoises went extinct and with them the principal easy source of sustenance for the settlers. The community was remote, farming was hard, water generally

Opposite: A native short-eared owl despatches one of the many introduced animals that breed in the highlands around human settlements.

GALÁPAGOS MURDER MYSTERY

It was in the wake of the abandoned Norwegian settlement that an odd pair, Dr Friedrich Ritter and Dore Strauch, came to stay. They were by turns married (to other partners) vegetarian philosophers with nudist tendencies. Ritter had his teeth removed as a precaution and used sturdy steel dentures when the need arose. Strauch later shared them when her own teeth were lost. They had read descriptions of Floreana and its history, and named their place Eden – perhaps they should have known better.

Friction began in this fledgling colony when the more conventional Wittmer family – Margret, Heinz and son – arrived. The two households did not hit it off. In fact, they hated each other, but bloodshed might have been avoided had it not been for the appearance of a third group. The self-proclaimed Baroness Eloise Wehrborn de Wagner-Bousquet pulled up to the island with three male consorts and a revolver, and futher declared herself 'Empress of Galápagos'. They named their resort Hacienda Paradiso. There are enough lurid details to fill a book (and it has been written – *The Galápagos Affair* by John Treherne – see Further Reading, page 233), but ultimately the body count says it all.

The baroness and her favoured lover Philippson disappeared without a trace under highly suspicious circumstances. No one was ever brought to trial. Lorenz, a former lover who had been systematically abused mentally and physically by the baroness and Philippson (and had sought shelter with the Wittmers), was found mummified on a distant island together with a surviving Norwegian settler who had unwisely offered him a lift. Ritter, the philosophical vegetarian with lapses, appeared to have something to say on the matter, but was poisoned by a bad chicken, or possibly by Strauch. She wrote that on his deathbed he seemed to say, 'I go; but promise me you will not forget what we have lived for.' Margret Wittmer, who was also there at the end, recalls that he wrote his last sentence to Strauch, 'I curse you with my dying breath.' Strauch and Margret never could agree.

Strauch left the islands and was killed in Nazi Germany during the Berlin bombing. Margret Wittmer lived on, on Floreana, running a guesthouse to the chelonian old age of 96. She and her island did eventually find a little peace, once their neighbours in 'Eden' and 'Paradise' had passed on. 'I like very much to live on Floreana alone,' she once confided to a visitor. And no one could blame her for that.

scarce and orchilla moss didn't make a profit. To escape penury people scattered to start a life on the other uninhabited islands. Those that remained overthrew Williams, who sailed away for ever. Perhaps only a dozen people were left by the time Villamil returned to inspect the ruins of his project.

MANUEL J. COBOS, ENTREPRENEUR, TYRANT

Because Galápagos was not at that time considered strategically situated and none of the aggressive powers really wanted it as a possession, it stayed with Ecuador almost by default. And Ecuador had enough on its plate on the mainland to worry too much about these distant islands.

The lack of government interest attracted lone entrepreneurs with ideas of building their own kingdoms. Manuel J. Cobos, an Ecuadorian, was a bullish and tyrannical example of the breed. In 1888 he started a settlement on Chatham Island called 'El Progreso' (progress) and filled it with convict labourers and desperate settlers. He planted sugar cane, vegetables and fruit trees – and for a while they prospered in the cleared volcanic soil far away from the mainland crop pests that followed later. Cobos's main success lay in addressing the islands' age-old water-shortage problem by diverting meagre fresh-water streams into irrigation for his farms. But his treatment of his workforce was akin to slavery, and payment if it came was in a currency of his own issue, valid only in his own shop. He flogged men to death and sent many others to a slower doom by abandoning them without provisions on desert islands. He was eventually shot, hacked, stabbed and bludgeoned to death by an expressive mob. The subsequent trial in Ecuador sentenced two men to prison, but concluded that everyone else had suffered quite enough. Cobos was buried with ghosts and bitter memories on the very site where he had earlier executed five men by firing squad. Progress was halted.

'ALL IS ARRANGED BY NATURE'

The final colonizing effort in our trio brings us more up to date. In 1926 a hardy contingent of Norwegians came to a still largely uninhabited Galápagos with tonnes of equipment and dreams of setting up a fish and sealing industry 'in paradise'. They were tough, resourceful and optimistic. Their leaders were no tyrants. With Galápagos waters so full of fish and seals, it is hard to see how they could fail. Norwegian newspapers described how 'all is arranged by nature so that an industrious and energetic colonist could be happy'.

The colony lasted just two years. You can find the remains of their deserted main settlement just behind the barrel of Post Office Bay today.

Opposite: The mummified body of Lorenz, who died as a castaway shortly after the mysterious disappearances on Floreana in the 1930s.

TO THE PRESENT

These stories are not some small sample from a larger uneventful or happy picture; they are the story of Galápagos again and again. It is hard not to narrate it all as tragic farce. Yet bereavement and suffering, the cost of broken dreams, were all too real.

Ironically, the Second World War drew a sort of line beneath those colonization attempts that ended in heavy human casualties. Construction of a military runway brought contact with the mainland and prevented the worst excesses of hate and tyranny – but still groups came here looking for a new life, and left (if they survived) finding that they preferred the old. A brutal prison colony on Isabela managed a little bloody anarchy as recently as 1958, with a three-day siege of Puerto Villamil to add to an unknown number of dead prisoners in a hellish camp.

All that can be said in a positive vein is that, were it not for repeated human failure, the islands would surely have been denuded of their wildlife long before the present day. It was Dore Strauch, of Galápagos murder-mystery fame (see box, page 50), who penned a simple conclusion echoed many times through centuries: 'These islands are, in truth, one of those places on earth where humans are not tolerated.'

Ultimately the lifting of the human curse on Galápagos came when it was realized that the islands offered a different sort of resource that could perhaps be used without destroying it. Darwin's visit over a hundred years before had resulted in the place being seen by the rest of the world as a monument to nature and the muse for humankind's most revealing idea. Scientists came in droves, at first to take 'the last' specimens from what were declared 'wondrous dying islands', but later just to observe. With the government of Ecuador they founded a national park and encouraged eco-tourism as a way the islands might be enjoyed without taking from them. It seemed as if the lesson of staying in harmony with Galápagos might have been learned, and finally some communities prospered. Tour boats began in 1969 and, by the mid-1980s tens of thousands of people from every nation were coming each year to witness what has often been called 'the last paradise'.

But even tourism has its limits, and the urge to harvest natural resources is very strong in our species. Today the old riders of misery and destruction are cantering back into town. Many devastating pests arrive with human cargo every year. Seemingly irresistible pressures to fish, farm and build are growing here together with a human population that has doubled every ten years and will soon exceed 30,000. Nearly 500 years since their discovery, and nature's great golden goose is ailing again. There seem to be no happy endings on Galápagos. In CHAPTER 7 we will find that our own times on the islands may yet prove the saddest of them all.

Opposite: National park guards fight to protect the surviving giant tortoises from feral animals and hungry fishermen on Isabela.

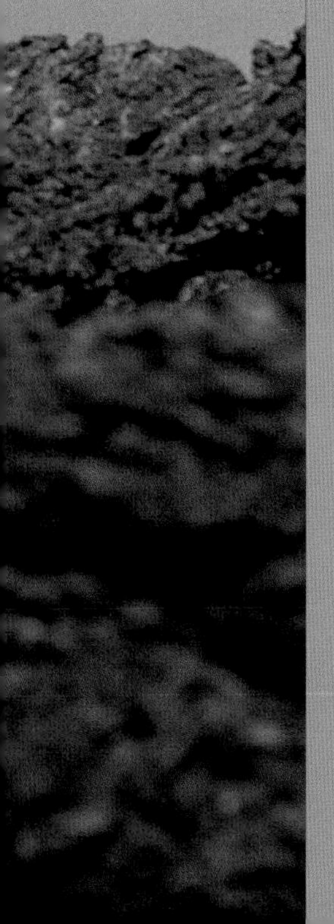

3

In 1859, Charles Darwin's *Origin of Species* brought a completely new view of life on Earth: species are not permanent, the perfect work of an intelligent creator, but are *continuously changing*, one form into another; what we see today is simply a snapshot of the ones that have *struggled* and *survived*, chosen by the *blind forces* of nature. Though in the book's 500 pages Galápagos is mentioned just a handful of times, even as the *Enchanted Isles* became a fading memory, a place once visited in a distant youth, Darwin held on to one certainty – that they were *the origin of all his views*: the origin of the *Origin of Species*.

DARWIN & EVOLUTION

FIRST IMPRESSIONS

Fit, impressionable, homesick: a 26-year-old Darwin first caught sight of Galápagos on 15 September 1835. 'A steady, gentle breeze of wind & gloomy sky,' he wrote in his diary. Captain FitzRoy nudged HMS *Beagle* closer and dropped anchor just north-east of the modern town of Puerto Baquerizo Moreno on the island of San Cristóbal, known then as Chatham. Everything was black: the beach was black lava, solidified, jagged and buckled, behind it, a low horizon of black, lifeless cones. The sweet anticipation of 'Land ahoy!' had been replaced by a parched, volcanic reality: first impressions were not favourable.

Previous page: A mockingbird picks ticks and dead skin from a land iguana. Above: A large ground finch and a small ground finch: two species separated by the design of their beaks.

But for Darwin, something was immediately different. Where others saw only menace, he quickly perceived more. For an Englishman, rock is the very symbol of permanence, but here the ground, though hard and metallic under hobnail boot, looked in places more like a liquid, like the sea 'petrified in its most boisterous moments'. Walking on these silvery swirls, ropes and waves of lava, frozen just yesterday, Darwin began to see a land not ancient and constant but in perpetual flux, changing by the year, by the day.

Darwin's England was at the height of industrial revolution and change was rife. But while studying at Cambridge for a career in the church he had been introduced to a very different world by the botanist John Henslow. With summer field trips and scientific soirées, Henslow imbued Darwin with a passion for the natural world which, in complete contrast to the industrial, was then seen to be unchanging and permanent, its beauty and complexity hard evidence of God's design. It was also Henslow who put Darwin in touch with Captain Robert FitzRoy.

FitzRoy was embarking on an exhaustive survey of the coast of South America and was looking for a gentleman-naturalist to accompany him, to collect rare specimens where appropriate, but more importantly to help look after his sanity. (This was a serious issue: in a fit of melancholy the previous captain had blown his brains out somewhere in the lonely wilds of Tierra del Fuego.) Through the Cambridge grapevine Henslow heard of FitzRoy's predicament and soon recommended his promising student. In those days, not many had a round-the-world trip under their belt. If Darwin took the opportunity seriously, there was a good chance he could turn himself into a qualified man of science – a far more exciting prospect than 'country parson'. After some reluctance from FitzRoy concerning the shape of Darwin's slightly bulbous nose, which he thought showed a lack of inner resolution, and further stallings from Darwin's father, who thought the whole expedition a waste of time, the matter was settled.

As a parting word, Henslow advised Darwin to read Charles Lyell's controversial new book, *Principles of Geology*, which was causing quite a stir in the scientific establishment. Henslow didn't agree with it, in fact he refuted much of it, but thought his protégé should at least be aware of it. He nervously added that Darwin should 'on no account accept the views therein advocated'.

LAND OF CRATERS

Principles of Geology described a physical world very different from Henslow's; a physical world whose features – mountains, rivers, coasts – were constantly evolving, shaped through time by forces that have the same effect today. No sooner had Darwin left England than the book came into its own. In Cape Verde he saw volcanic islands of recent origin and

Nothing could be less inviting than the first appearance ...
The black rocks heated by the rays of the Vertical sun like a stove, give to the air a close & sultry feeling ...
The country was compared to what we might imagine the cultivated parts of the Infernal regions to be.

Charles Darwin

in mainland South America more volcanoes – one erupting – fossils and uplifted mountains. He was caught in an earthquake: 'the world, the very emblem of all that is solid, moves beneath our feet like a crust over a fluid'. Now, from mariners' stories of a land of fire and brimstone, he was looking forward to Galápagos as much as any part of the voyage. He was more than ready to observe geology in action.

Disappointingly the great volcanoes were sleeping. But Darwin did see, from the lava that had spewed from their mouths, from those solid waves underfoot, that many had been active in the very recent past. He noticed that other volcanoes were composed of sandstone-like 'tuff' and had their southern sides 'quite broken down and removed'. He was the first person to work out that these must have been formed in the sea where the prevailing swell from the southerly trade winds had worn them down. Darwin was beginning to get the picture:

Above: Not the familiar grey-bearded sage, Darwin was a young man when he visited the Galápagos and is shown here shortly after his return.

> *Seeing every height crowned with its crater, and the boundaries of most of the lava-streams still distinct, we are led to believe that within a period geologically recent the unbroken ocean was here spread out.*

This was Darwin's first big Galápagos insight. Here was new land, born hot and sterile from the depths of the ocean. Life had come later.

Thanks to Lyell and to his own recent travels, in Darwin's mind these first footsteps on Galápagos were quickly becoming the catalyst for a chain reaction of thought that once in motion would not be stopped, that would define his life from this point on, that would scare him rigid, and that over the next 25 years would lead him in steps through further remarkable insights to possibly the greatest thought of all time.

COLLECTING FRENZY

On the third day in Galápagos, Captain FitzRoy moved the *Beagle* to Stephen's Bay, also on San Cristóbal, anchoring near present-day Puerto Grande. Here Darwin first saw marine iguanas.

> *The black Lava rocks on the beach are frequented by large (2–3 ft) most disgusting, clumsy Lizards. They are as black as the porous rocks over which they crawl & seek their prey from the Sea—Somebody calls them 'imps of darkness'.*

How appropriate that these lizards, like shards of rock themselves, marked a shift of Darwin's attention from geology to wildlife. They lived on the land and fed in the sea – but on what? He dissected several on the *Beagle*'s chart table to find out. A belly full of seaweed. He'd never seen anything like it and was further astounded by their numbers.

Above: Darwin first described the crater-pocked Galápagos landscape as being like the iron foundries of Staffordshire.

And what of the famous tortoises? Only three decades after whalers started using Galápagos as a base, hunting for the ship's larder had taken its toll. FitzRoy had the crew scouring for them but initially found none. Darwin noticed some tracks, well worn by the regular passage of weighty feet. Soon he met them face to face.

He upturned them to see if they could right themselves; they could. He rode them and marvelled at their strength. He measured their speed: 30 yards in 5 minutes – that's 6 km in a day – quite fast enough to get where you need to go when your world is a small island. But though Darwin was by now completely sold on geology in action – on Galápagos, the brave new world – he still saw these lumbering giants as creatures from a land that time forgot. 'Surrounded by black Lava, the leafless shrubs & large Cacti they appeared most old-fashioned antediluvian animals; or rather inhabitants of some other planet.'

Previous page: 'Imps of darkness' – marine iguanas – soak up the equatorial sun with a view of Fernandina. Above: Marine iguanas usually dive for ten minutes or so but they can stay underwater for as long as an hour – as one of the Beagle's *crew discovered when he tried unsuccessfully to kill one by tying it to a weight and sinking it.*

As FitzRoy moved the *Beagle* methodically onwards in his effort to chart the islands, Darwin took every opportunity to go ashore. And as he followed the wide tortoise paths into the interior with his servant Covington, he began to collect in a frenzy. He skinned birds and netted insects. He pickled fish, reptiles and snails. He pressed hundreds of plants. But on these islands he found just one native mammal – a mouse (we now call it a rice rat). It seemed that in this strange cast of characters, all the parts usually played by mammals were taken by reptiles. And there was something else about the animals, especially the birds: Midshipman King caught a dove in his hat; a mockingbird alighted on the edge of Darwin's cup and sipped the water as he lifted it off the ground; he pushed a hawk off a tree with the barrel of his gun. It was almost eerie: all the animals seemed caught in an Eden-like state of fearless innocence.

HAVEN OF PEACE

Just three years before Darwin's arrival, the famous post barrel at Post Office Bay on Floreana had become the focal point for the first human settlement in Galápagos, called Haven of Peace (see CHAPTER 2). Here, Darwin and FitzRoy were met on the beach by the Acting Governor of Galápagos, conveniently the Englishman Nicholas Lawson. He led them away from the coast and up into the cool hills behind. The effect on Darwin was immediate:

> *Passing round the side of the highest hill; the body is cooled by the fine Southerly trade wind & the eye refreshed by a plain green as England in the Spring time.*

Relief at last from the heat and the thirst, if mixed with the pang of a glimpse of home. Floreana is high enough to catch rain but eroded enough to have a good soil, so here there was a 'tolerably luxuriant vegetation', a softer side to Galápagos and a wealth of new specimens. These islands were not just lava-strewn deserts, they were miniature worlds. Darwin collected everything he could, his curiosity now fully piqued. If he were right and Galápagos had been born lifeless, then from where had these extraordinary plants and animals come? Darwin was a Christian and a creationist so there were only two possibilities: either God had created them here or they had come from a previous creation elsewhere. 'I certainly recognize S. America in Ornithology', he wrote in his pocket notebook. This land seemed too new to have its own special creation, so in his burgeoning collections it made sense for him to search for similarities to life he had seen on mainland South America, for clues to a provenance outside Galápagos.

Perhaps that is why he paid very little attention to a rather remarkable piece of information. Lawson, through necessity, had developed a keen interest in the archipelago's wildlife and its uses to humans: in particular the differences between the islands. He told Darwin and FitzRoy with great delight that the Spaniards could tell from which island a tortoise came simply by the shape of its shell.

GARDEN OF EDEN

West of Floreana the islands get higher, blacker, meaner. As the *Beagle* rounded the southern end of Isabela, steam could be seen issuing from the crater we now call Sierra Negra. But FitzRoy's schedule gave no chance to climb the great volcanoes, so Darwin covered 36 pages of his notebooks with meticulous detail on crater formation instead.

The only notes of any length on wildlife relate to a new discovery: a second lizard, every bit as hideous as the first. This time it was orange, red and yellow, foraged on dry plants and up trees and lived in burrows in the ground. A land iguana. It had a facial expression that resulted in a 'singularly stupid appearance' and was just as

tame as the marine variety, so the temptation to tease it was great. Darwin waited till one was half buried in a burrow before pulling its tail. 'At this it was greatly astonished and soon shuffled up to see what was the matter; and then stared at me in the face, as much as to say, "What made you pull my tail?"'

Then came James Island, now called Santiago. Darwin, the surgeon Bynoe and their servants were put ashore at Buccaneer Cove while the *Beagle*, in serious need of water, returned to San Cristóbal to get some. For nine days they would be entirely marooned.

In the old days, Buccaneer Cove had been *the* popular anchorage for pirates, as the steep beach allowed them to repair their ships. Now the landing party met not pirates but a group of tortoise hunters sent from Floreana by their new friend, Lawson. Darwin quickly employed these men to fetch water from the nearby 'miserable little Spring' but, as a reminder of the vulnerability of man in this harsh world, this supply was soon swamped by a large wave. Had it not been for a passing American whaler giving them three casks of water, things might have turned out very differently for the world. As it was, Darwin and his companions had the run of the island, venturing deep into the interior, camping at the tortoise hunters' hovels and living on fried tortoise meat. As Darwin went he collected a flora and fauna virtually unknown in scientific circles, and here was his best chance to collect birds, particularly the smaller kinds now known as Darwin's finches (see box, page 81). In the highlands, as on Floreana, he found lush forests draped in long, wispy lichens, dripping with condensed vapour from the clouds; this time, at almost 1000 m, protected from the worst ravages of mariners and colonists, he also found springs swarming with tortoises.

Opposite: No one knows how long giant tortoises live in the wild. In his diary, Darwin notes that one was found in 1830 with the date 1786 carved by sailors on its shell – 44 years and still going strong (it required six men to lift it). Above: Darwin collected many species of finch but he never labelled the islands from which they came.

> *It was a curious spectacle to behold many of these huge creatures, one set eagerly travelling onwards with outstretched necks, and another set returning, after having drunk their fill. When the tortoise arrives at the spring, quite regardless of any spectator, he buries his head in the water above his eyes, and greedily swallows great mouthfuls, at the rate of about ten in a minute.*

Where others before him had seen nothing but a 'Hell on Earth', Darwin had stumbled on a Garden of Eden. But it was over almost as soon as it had begun. After only five weeks in the islands, at sunset on 20 October 1835, the *Beagle* set sail for Tahiti and then home. Though Darwin had seen many strange and wonderful things in his short time here, contrary to popular myth, in Galápagos there was no legendary moment of revelation. But the enchanted islands had cast their spell.

THE TORTOISE AND THE MOCKINGBIRD

In spite of his usual seasickness and by now desperate longing for home, no sooner had the *Beagle* turned her back on the Galápagos than Darwin went about the business of organizing his specimens. Ironically it was not 'Darwin's finches' that first caught his eye, but another very ordinary-looking bird, the same bird that had so cheekily alighted on his cup, the mockingbird. Looking at his specimens side by side, something now struck him as odd:

> *I have specimens from four of the larger Islands ... The specimens from Chatham and Albemarle Isd. appear to be the same; but the other two are different. In each Isld. each kind is* exclusively *found.*

Different islands within sight of one another each had slightly different-looking birds – in their plumage, but most noticeably in the shape of their beak. He then remembered the conversation on Floreana with Lawson: 'From the form of the body, shape of the scales & general size, the Spaniards can at once pronounce, from which Island any Tortoise may have been brought.' Again the same pattern: different islands, different tortoises. If Darwin were right and these animals had originally come from somewhere other than Galápagos, it seemed that on different islands of the archipelago mockingbirds and tortoises had then changed into different forms. As a creationist, Darwin believed that species were fixed in nature, as created by God. Even Charles Lyell, for all his controversial views on geology, upheld the sanctity of species. He believed that one type of animal or plant could vary to an extent but could not under any circumstances change into another. But if these mockingbirds in front of Darwin were different species, what then? He wrote in his notebook, 'such facts would undermine the stability of Species.' Had he caught a fleeting glimpse of evolution? It was too much to contemplate, so he concluded that they must be merely varieties. With a parting lament he no doubt returned to more comfortable thoughts of home: 'It is the fate of every voyager, when he has just discovered what object in any place is more particularly worthy of his attention, to be hurried from it.' His specimens would have to tell their own tale.

Left: A cactus finch and a carpenter bee both feeding on, and helping pollinate, an opuntia cactus.

THE BIG MISTAKE

Darwin reached England on 2 October 1836, five years after leaving home and one year after leaving Galápagos. He gave all his birds to the eminent ornithologist John Gould to see what he could make of them. Within a week, Gould presented his findings to the Zoological Society of London. The official meeting report reads:

> *a series of* Ground Finches, *so peculiar in form that he was induced to regard them as constituting an entirely new group, containing 13 species, and appearing to be strictly confined to the Galápagos Islands ... their principle peculiarity consisted in the bill presenting several distinct modifications of form.*

Darwin was dumbstruck. Yes, he had collected some finches in Galápagos, but most of the birds named by Gould as finches he had identified differently, as the European birds that they most resembled. He had labelled cactus finches as blackbirds and what we now know as the warbler finch he had called a wren. Come to think of it, he had been puzzled in the field as to why all these different birds had such similar plumage; now it was beginning to make sense. The mockingbirds, the tortoises – were these different species of finch also confined to specific islands? But with such heady excitement came the trough of despair as he realized his mistake. Although he had been meticulous in preserving his finch specimens for the journey home, he had neglected to label from which islands they came. It simply hadn't occurred to him.

Darwin was now convinced that out of one finch had evolved 13 different species and that each of those types had come about because they were somehow isolated from each other. It clearly wasn't as simple as the mockingbird or tortoise situation, since there were 13 species of finch and he'd visited only four different islands, but without isolation they would surely have interbred and merged into one type? The horrible realization dawned: these little brown birds could hold the key to how species came about, but without knowing which bird came from which island, he had nothing.

Frantically Darwin approached his shipmates for evidence. Both FitzRoy and Covington had collected birds, and they had labelled them as he had not. Lists of names circulated from Gould to FitzRoy and back to Darwin covered in question marks, but little could be salvaged: there was no proof that each species was limited to one island. To add salt to the wound Darwin had made a similar blunder with the tortoises. Out of 45 adults brought aboard the *Beagle*, none remained: they had all been eaten and their uniquely shaped shells thrown overboard.

Darwin looked to the mockingbirds for salvation. On this matter Gould was triumphant. Having pronounced almost all of Darwin's other 26 species of land bird as unique to Galápagos, he crowned it all by confirming that the mockingbirds were indeed three separate species.

THE 'ORIGIN OF ALL MY VIEWS'

Opposite: It was the mockingbirds, not the finches, that were the catalyst for Darwin's revelation. This is a Chatham mockingbird from San Cristóbal – collected by Charles Darwin, named and illustrated by John Gould.

It was coming together. Inspired by some drab-looking birds from a faraway land, Darwin opened his first notebook, called notebook 'b', on the 'transmutation of species', identifying 'species on Galápagos Archipelago' as the 'origin of all my views'. From his geological observations Darwin believed that Galápagos had risen from the ocean in the recent geological past and had never been connected to a continent. He had suspected that all life on the archipelago had come from elsewhere and that once there, in isolation on different islands, it had changed to form separate species. The finches were complicated, but the mockingbirds seemed to support this. Then came what was for Darwin the final confirmation: the distinguished botanist Joseph Hooker's report on his plant specimens. Unlike

GIANT TORTOISES

Since Darwin first heard from Lawson that there were different giant tortoises on different islands, they have been one of the most quoted examples of evolution in Galápagos. They still offer probably the clearest example of how different conditions on different islands can favour different traits. From an original 15, there are 11 surviving populations today, each a distinct type, some domed, some saddle-backed. (It is still debated as to whether they have diverged enough to be called species, hence some say 'race' or 'subspecies'.) As Darwin was told, most types are confined to separate islands, though it's not just isolation on islands that has caused their evolution. On Isabela there are five types of tortoise on as many volcanoes, separated not by water but by impassable lava fields. Oddly, most have domed shells but the one on Wolf volcano is a saddle-back. Some think it was brought there by sailors. Captain FitzRoy of HMS *Beagle* believed that all giant tortoises had been brought to Galápagos by sailors from the Indian Ocean. This might help to explain Darwin's astonishing failure to collect even a single adult specimen, thinking them not native to Galápagos and therefore of no importance. The *Beagle* did bring back four juveniles, but differences between shells are only noticeable when the tortoises grow old, and that takes a very long time. Many live for 100 years and some estimates don't rule out there being a tortoise alive today that could have met Darwin.

A far better insight into how tortoises probably reached Galápagos comes from a collecting mission organized by the California Academy of Science in 1905–6. A rowing boat carrying two tortoises capsized and was smashed on the rocks. The crew had to return to their ship by land. The next day they saw the tortoises happily bobbing about like corks. They had been in the water for 18 hours and were still going strong.

with the finches, Darwin had fortuitously labelled many of his plants by island; conclusions about these could give proper insight. Out of over 200 species identified by Hooker, not only was an incredible half unique to the Galápagos but also, most importantly, three-quarters of those were confined to single islands and all were 'out and out American' in type. As more reports of the unique nature of his collections kept coming, all supporting the 'truly wonderful fact' that different islands had different species and all bearing marked similarity to animals from the American continent, Darwin's suspicions were confirmed. He began to see through time:

> *I fancied myself brought near to the very act of creation. I often asked myself how these many peculiar animals and plants had been produced: the simplest answer seemed to be that the inhabitants of the several islands had descended from each other, undergoing modification in the course of their descent; and that all the inhabitants of the archipelago were descended from those of the nearest land, namely America.*

What Charles Lyell had done for the inorganic world, Darwin could now do for the organic. Why shouldn't life, like geology, be constantly changing, with the present linked to the past by a set of constant rules? In South America, Darwin had personally unearthed the fossils of giant armadillos and sloths exactly where their diminutive relatives roam today, the modern and the ancient types separated only by time. In Galápagos he'd then seen related living species separated in space, in isolation from island to island. In his mind's eye he now saw all species past and present connected through both space and time on a vast tree of life. To Darwin, evolution was fact. What he needed now was a mechanism to drive it.

SURVIVAL OF THE FITTEST

In the autumn of 1838 Darwin read *An Essay on the Principle of Population* by the economist Thomas Malthus. Malthus argued that human populations had a natural tendency to reproduce at an unsustainable rate but that it was famine, epidemic and war that checked that extra growth and kept populations stable. This was just what Darwin was looking for. He had applied Lyell's rules of geology to biology, now he borrowed from economics; again he added his own insight, realizing that those that died in the struggle for life would be the weakest and those that survived the strongest, the best adapted, the fittest.

It's easy to see how all this relates to Galápagos. It makes sense that the descendants of animals that could best survive the crossing of an ocean should prosper in such a harsh place. And in a land freshly emerged from that ocean, there was initially no competition, so reptiles came to rule where there were no mammals, and finches took on the roles of other missing birds. Perhaps, from island to island, where conditions were subtly different, a slightly different beak or a different shell gave an advantage; then that was the 'fittest' that survived. It even explained the lack of fear in the birds: with no humans to avoid, there was simply no need of it. One world's weakness is another world's strength.

But for Darwin, 'Malthus in Galápagos' could only ever be a pipe dream. It was hardly a question of hopping on the next long-haul flight to test his theory on the other side of the world. So as the islands sank back into obscurity under a tropical sun, Darwin retreated into his thoughts. His startling collections had done him proud and he was fast becoming a man of scientific standing; his hero Lyell had adopted him as his own. But his ideas about evolution denied the need for God and in a society that saw the work of God in all life, revealing those thoughts would court disaster. His wife, Emma, was already concerned for his salvation. He confided in his friend Hooker that it would be like 'admitting to a murder'.

So he bided his time, looking to the world around him, for 20 years gathering evidence for evolution from dog breeders, pigeon fanciers, farmers, barnacles, embryos, fossils and even his hairdresser. It was going to take more than a few notes from Galápagos to convince the world of this heresy.

THE BIG IDEA

On the Origin of Species by Means of Natural Selection finally went on sale to the public on 24 November 1859 at a price of 15 shillings. All 1250 copies sold on the first day.

It caused a sensation. Overnight it whipped up a frenzy of interest and changed the world irreversibly. Ironically, although Galápagos featured only fleetingly in the book, what Darwin came up with reads like a page from the Galápagos tales of old, with the environment acting like a deranged Old Testament God, selecting the traits that best suited survival by culling the weak and pardoning the strong. But this god, called 'natural selection', had absolutely no direction or purpose: 'There seems to be no more design in the variability of organic beings and in the action of natural selection than in the course which the wind blows.'

With the publication of *Origin of Species*, the journey of understanding set in motion on a five-week brush with the Enchanted Isles reached its natural conclusion. Though Darwin's research and writing were prolific till the end, not able to return to Galápagos and without the help of modern genetics, this was as far as he could take it. It was enough for one life. But for evolution, it was just

Opposite: A blue-footed booby 'sky-pointing' as part of its courtship display. Darwin would ultimately see such sexually selected behaviour as an integral part of man's own evolution.

the beginning. Though *Origins* marshals an enormous weight of evidence that evolution has happened, nowhere does it actually document the origin of a species. Darwin had thrown down the gauntlet and Galápagos had become *the* place to put evolution to the test.

LOOKING FOR EVOLUTION

Throughout the late 1800s and early 1900s, a Galápagos craze raged. Expeditions went there looking for evolution or not, depending on what they believed, and all came home with specimens to fill the temples of Darwin's new religion: museums. Here they were methodically measured, catalogued and named, and among them were many new species, especially as the expeditions moved to less visited islands. But in spite of a great deal of effort, no one got much closer to finding out what was really going on. As for the islands, it nearly killed them. One expedition alone brought back 3000 birds and 65 live tortoises and another an extraordinary 75,000 specimens, including 264 tortoises and 10,000 birds. Even the sole tortoise on Fernandina was killed and skinned in the name of science. Not until science learned to study Galápagos in Galápagos would any real progress be made.

In 1959, exactly 100 years after the publication of *Origins*, UNESCO created the Charles Darwin Foundation for the Galápagos Islands in order to organize and maintain a scientific-research centre there. The Ecuadorian government established the Galápagos National Park and the following year the first buildings of the Darwin Station appeared in Academy Bay on Santa Cruz. The very qualities that make the islands so good at evolving species make them an equally good place to study those species evolving. Isolated islands born virgin from the ocean, then populated by creatures from elsewhere, the tameness of the animals and different casts of characters on different islands subject to different conditions; the whole at the mercy of ocean currents that bring drought and deluge on a regular basis. To science Galápagos is like a dozen giant Petri dishes, each with its own set of different experimental controls. Here, over the years, every aspect of Darwin's thinking has been gone over with a fine-tooth comb and all the time new discoveries are being made: seven species of lava lizard, 15 scalesias (relatives of the sunflowers), a remarkable 65 species of snail, not just on different islands but on different patches of vegetation separated by lava; blind fish that have evolved from the sea to live inland and underground in dark lava caves. With DNA studies, evolution can now be looked for not just at the species level but also in the genes. Geologists have recently even found new Galápagos islands: in fact old islands that have lived their life and sunk beneath the waves. The more we look, the more remarkable this place becomes. But there is one discovery

DARWIN'S FINCHES

Darwin was always half suspicious that he had seen similar finch species living together; he was later found to be right. But he was also sure that isolation was the key to speciation and that if similar types lived together they would simply merge. Fifty years after Darwin's death, the mystery still hadn't been solved. The latest theory blamed rampant breeding among species for such a wide range of finches – a hybrid swarm – the resulting different beaks of no significance. In 1938, David Lack, a 25-year-old schoolmaster, took up the challenge of going to Galápagos to find out what was really going on. His living conditions were basic. His companion nearly died of dysentery. When the breeding season came to an end and he had not seen a single between-species mating, he concluded that the finches were not hybrid swarms, but because he had seen different finches apparently eating the same food, he had to agree that the different beaks had no adapative significance. Fed up with Galápagos, he went home via America to look at the now famous collections of Galápagos finches at the California Academy of Science and the American Museum of Natural History in New York. He measured 8000 finch beaks. Then, just like Darwin, several years after he'd left Galápagos, he had his inspiration. He noticed that the species of finch whose beaks are most nearly identical do not live on the same island.

Galápagos finches were most likely one of the islands' earliest colonizers, since they seem to have had the pick of the habitats. Some remained feeding on the ground, some took to the trees and others evolved to feed like woodpeckers or warblers. On different islands different types prevailed: so far, just like mockingbirds and tortoises. But these birds can island-hop and that's where things get interesting. If they come together, they become competitors and, as Darwin said, only the fittest will survive. They can live together only if their slightly different beaks mean they can avoid competition by eating slightly different foods. And that was Lack's solution: the birds are isolated not on different islands but by different food. He speculated that it is in the dry season when food is scarce that the beak shape becomes critical and that's why he hadn't seen it in Galápagos: he was there when it was wet. He was later proved right. Lack called his book *Darwin's Finches* and the name stuck.

Left: It was only after returning home that Darwin realized the true significance of the finches. In this illustration from the Beagle*'s voyage he starts to group them together and remarks that 'one might really fancy … one species had been taken and modified for different ends'. (1. Large ground finch. 2. Medium ground finch. 3. Medium tree finch. 4. Warbler finch.).*

that stands above all others because it finally brought Darwin's Galápagos journey of evolution full circle.

The island of Daphne Major is nothing but a crater, rising steep sided from the ocean. It is tiny, there's no beach, only one place to land and one place to camp: a difficult place to live but perfect for a study of a confined population. In 1973, Peter and Rosemary Grant went to Daphne to study the birds that so perplexed Darwin: the finches. There are four species on Daphne, and the Grants started catching the tame birds, measuring their beaks and then releasing them. They were hoping to solve, once and for all, the mystery of how the finches got their different beaks.

For five years, not much out of the ordinary happened. But in Galápagos when drought comes it is severe and in 1977 it was biblical. Plants withered and seeds vanished; finches died. Probably the worst hit was the medium ground finch, which did not breed at all and suffered an incredible 85 per cent decrease in population. But most interesting to the Grants was that it was not any old finch that died. Medium ground finches eat seeds, and their beaks are sharp at the tip and stout at the base: perfect for probing, picking and cracking. But in every population some birds' beaks are larger than others (just as there are people with larger or smaller noses, like FitzRoy and Darwin). There are also larger and smaller seeds. In the severe drought the finches were all struggling to find food and the smaller, softer seeds were soon all eaten up; only larger, harder seeds were now left. Birds with big beaks to crack hard seeds survived; birds with small beaks died. In the next breeding season, when the rains returned, the big-beaked birds bred and passed on their big-beak genes to the next generation; by 1978 the average size of a medium ground finch's beak had grown.

MODEST SUCCESSES

And the meek shall inherit the earth. Well, certainly on Galápagos this seems to be true. Look for the species that diversified most on the islands and you won't find the globe-trotting owls or raptorial hawks, but rather the finches, snails, tortoises, scalesias and mockingbirds. None of them on the face of it the sort of dynamic founding fathers you might expect to see generating influential dynasties on a new frontier. And paradoxically it's their very unadventurousness that leads them to such success.

Species like hawks, owls and killer whales travel relatively easily between the Galápagos islands and the mainland. But if you are a snail or a finch, not only is it something of a miracle that you made it here at all, it is even tough to get between adjacent islands. Hence it is quite easy for a population to find itself isolated, and isolation is the first step along the road to developing a new species. Without frequent visits from members of the main population, evolution is free to follow its own independent course. It's another Galápagos insight into the origin of species.

Above: The flightless cormorant is one of the delights of Galápagos evolution that Darwin never saw. It has sacrificed the power of flight in favour of a more streamlined body shape for hunting in the water.

Then, in the El Niño year of 1982–3, the pendulum swung back. Instead of the usual couple of months' breeding season, there were eight. There were seeds and birds everywhere. But this time, small, soft seeds predominated: a paradise for birds with smaller beaks. Bird numbers rose and outstripped the island's resources, but it was big-beaked birds eating big seeds that suffered most. The average beak size got smaller. On Daphne Major, in the space of just a few years, the Grants had seen beaks get bigger and then smaller in front of their very eyes. At last evolution by natural selection had been seen in action in Galápagos. The results caused the usual debate, and still no one had seen a new species actually come into being. But further work calculated that at that rate of change, a new species could evolve every 200 years. Evolution's biggest problem has always been the length of time needed for change to come, so predictions like that make it worth looking further. And the fact remains: the place where Charles Darwin, quoted below, first caught a glimpse of it will always be the best place to find it:

> *Hence, both in space and time, we seem to be brought somewhat near to that great fact – that mystery of mysteries – the first appearance of new beings on this earth.*

4

The various Galápagos islands have been described as being like members of a large family, all similar, but each with its *individual quirks and oddities.* Foremost among those oddities must be the plants and animals that inhabit them. They are creatures that have *amazed, enlightened* and even *appalled* people since their first discovery. Just their names evoke curiosity – *giant tortoises, lava lizards* and *cactus trees*, to name but a few. And the habitats they occupy are no less diverse, from boggy moorland to elfin woodland, active volcanic craters and desert island scrub.

LIFE ON LAND

FIRST IMPRESSIONS

Standing on the high lava fields of Wolf volcano, you can witness giant tortoises, some as heavy as a pony, trekking cinder slopes among impassive cliques of golden land iguanas. Lava trails have been worn smooth by countless generations of scaly feet as these reptiles pass on their way to drink at the damp ferns and lichens that catch the morning mists. It is like some archaic scene from a lost world, but it is in fact just an alternative path to the present. A world where large mammals never naturally settled, where the key herbivores are reptiles, and the top predators are birds and snakes. It is a strange land both distant and unfamiliar.

Previous page: Giant tortoises have roamed the volcanic calderas of Galápagos for more than a million years. Above: In hot lava colours, the land iguanas flow from their night dens on the upper reaches of Fernandina's volcano to seek the warmth of the morning sun.

ISLAND CASTAWAYS

Distance and unfamiliarity were probably the two major factors in making these islands' creatures strange in the first place. In CHAPTER 1 we saw that every animal or plant to arrive on Galápagos had jumped one very large hurdle – about 1000 km of open ocean. They came on rafts, flew or floated to the islands, a journey long and hazardous enough to weed out the large mammals and many plants. The habitat that greeted survivors was probably quite unlike the one they had left at home.

Iguanas on the mainland enjoy lush forests and fresh waterways – they must have been unusually adaptable to survive on Galápagos. Plants such as orchids require particular pollinators to be present, but one partner cannot survive without the other, leading to a colonization catch-22 that many species share.

The place is like a new creation; the birds and beasts do not get out of our way ... and all this amidst volcanoes which are burning round us on either hand. Altogether it is as wild and desolate a scene as imagination can picture.

Lord George Anson Byron, *Voyages of HMS Blonde* (1826)

In fact the rules that determine which, and how many, plants and animals settle on an island have a certain natural geometry. The broad principles are that – all else being equal – smaller islands hold fewer species than larger ones, and distant islands hold fewer species than those near a source of potential colonists. In fact all else is not quite equal on Galápagos. Islands with tall volcanoes receive more rainfall and have larger temperature gradients, suiting a broader range of species' needs. Their welcome mat is bigger.

Despite the trials of arrival, over 560 species of native plants, over 55 species of native land vertebrates and 1700 species of native insect are found on Galápagos. Of course not all these species 'arrived' in the strict sense. Once on the islands, given time and genetic isolation, they changed and diverged to the point where they could no longer be considered the same species as those that first clambered up the beach or slumped exhausted on a thorny perch. One species may split and split again, to give rise to new species, as happened here for the finches we met in Chapter 3. Such evolutionary change and 'radiation' is what made Galápagos animals unique. When biologists talk about 'adaptive radiation' they are referring to the way a species, finding itself in an underpopulated 'land of opportunity', can sometimes adapt into available, unfilled niches. When finches arrived on Galápagos there were no woodpeckers, few grazers, not even many flower visitors. These vacancies allowed the original ancestral finch to split into different species that probed for insects in wood, ate leaves and specialized in cactus flowers.

KEY HABITATS

Despite the important differences between islands, we can identify three broad categories of habitat, each with its distinctive species. They are: the coastal zone, the dry lowlands and the moist uplands. Using the diverse major island of Santa Cruz as a template, these zones can be further divided into seven habitats that have

ON BEING UNIQUE

An endemic is a species of plant or animal that is found in a particular place and nowhere else. The Galápagos has many such species. Before humans arrived nearly half the plants and more than 70 per cent of the land vertebrates were unique to these islands. Not every plant or animal here is found throughout the archipelago, however; each island has its own cast of plants and animals specific to particular habitats and assembled, or evolved, from the chance arrivals that happened upon their shores. Nearly 20 per cent of Galápagos land species are unique to just one of the hundred or more vegetated islands and islets. They often have close relatives on nearby islands and more distant relatives on the South American mainland. As we saw in CHAPTER 3, it was this observation that so impressed Darwin and caused him first to wonder how it could all be explained.

Darwin's conclusion, reached years later, was that when a group of animals is isolated from the parent population their 'design' gradually changes. We now understand that this is the result of their changing genetic constitution, the recipe for that design. This change is largely effected through natural selection, favouring those whose heritable traits, arising from natural variation, fortuitously adapts them to local circumstances. Eventually the colonists become very different from the parent population and can be called a 'new species' when they will no longer freely inter-breed with the parent population even if their ranges once again overlap.

parallels on various other islands. To see them all we might travel, as a castaway or naturalist would, from the coast to the highlands. The rich coastal zone is the theme of CHAPTER 5. Here we shall start just beyond its cliffs, beaches and mangroves, in an altogether more hostile world.

THE LAVA FIELDS – A BRAVE NEW WORLD

Barren lava fields dominate the coasts of many of the younger islands. Hardly inviting, but certainly compelling, they are the end product of molten-rock rivers that have gushed from vents perhaps hundreds of metres up from the coast. Once on level ground these rivers spread out as burning estuaries of lava, which cauterize the old surface and landscape it anew. The most recent flows are forbidding places, apparently devoid of greenery and shimmering with the sun's heat almost from dawn. Away from the coastal zone, these lava fields can look quite dead. Water evaporates or sinks quickly through the permeable rock, leaving little time for pools to form. Walking across the roughest of them is a near impossibility for most large animals – the lava forms barriers to travel that isolate land communities as effectively as the sea. It is this habitat that welcomed the very first colonizers to these islands and, despite all appearances, creatures can live here.

If you arrive in the morning you can see the webs and threads of spiders almost everywhere. They are among the first creatures to reach any new land. As tiny hatchlings they can parachute vast distances, carried by the wind on silken threads. They will feed on other accidental arrivals or on the inconspicuous micro-grazers that have found the equally tiny algae, lichens, bryophytes and ferns that have already settled among the lava. Where there is moisture in the air, the lava can become quite colourful with the red, yellow, green and grey bodies of lichens and algae. These ancient communities of plants and animals survive here by virtue of being physiologically capable of toughing out conditions with which few 'modern' organisms can cope. They are the vanguard of colonization on these islands and another glimpse into a universal past.

Above: Cactus – spines and all – offer a feast for land iguanas.

Silently and slowly these simple pioneers erode the rocks, absorb nutrients and form tiny pockets of soil that will allow other plants and animals to join and eventually replace them. An example of these secondary colonizers is the straggly-looking little endemic herb, *Mollugo*. Looks can be deceptive, of course, and though feeble above, its roots pierce and fracture the lava rock below, softening the landscape further. Elsewhere on Galápagos gloves of endemic *Brachycereus* cactus grow in the bare lava, their distinctive orange stem-tips almost incandescent in the dawn. A dense coat of spines protects these cactuses, ensuring that the water they hoard will not easily be surrendered to any passing herbivore. And such creatures do venture here.

A LOST WORLD

If this phrase means anything, then it is here on Galápagos. There is evidence that the very first Galápagos islands may have emerged and disappeared during the Cretaceous – the age of the ruling reptiles. If so, then half-drowned dinosaurs may once have found salvation on Galápagos shores, ichthyosaurs may have swum in their underwater caves and skin-winged pterosaurs soared among the lava cliffs. It is intriguing to wonder whether evolution in isolation once made unique Galápagos dinosaurs or encouraged a finch-like diversity among the flocks of little pterosaurs. It's possible, but we will probably never know for sure. Sadly the earlier Galápagos islands, and any fossils of creatures that occupied them, have long since been dragged beneath the ocean. There seems to be no biological continuity between those islands and the ones we see today.

VISITOR FROM ANOTHER WORLD

There is a species on Galápagos so little known that many guides deny its existence. This is *Ogilbia galapagosensis*, a creature that has only ever been found on a few occasions in the openings to lava tunnels and crevices that hold fresh and brackish water. It is a classic cavefish, unpigmented and nearly eyeless, adapted for a life in darkness. When little, white Galápagos cave shrimps pass it by there is an explosive gulp from deceptively large jaws and it swallows them whole. *Ogilbia* is a representative life form from the unknown but extensive ecosystem of interconnected water-filled lava tubes and crevices that permeates deep into these islands. Its ancestor may have been close to a bright orange relative, *Ogilbia deroyi*, discovered by the DeRoys, in the 1960s. That fish still inhabits the coastal reef and likes the darkness and cover of rocks. It would have been only a small move for a common ancestor to enter the cave system and there lose its eyes and pigment by the forces Darwin outlined for troglodytes in *Origin of Species*.

Land iguanas walk on to bare lava fields for the scant promise of food. Their orange and red skins are a remarkable contrast to the blacks and greys of cooled lava. Reptiles have an advantage here over mammals, in that their sealed, scaly skins and highly efficient kidneys release little water back into the environment. Their dependence on ambient heat to keep them at operating temperature saves energy and makes them happy with surprisingly meagre meals. The price they pay for this low-demand lifestyle is that they need some warmth in the morning just to get moving, and must seek shade rather than sweat it out during the middle of the day (though, as we shall see in Chapter 5, marine iguanas are rather a special case). Fortunately the lava offers many small niches and caves that land iguanas and their smaller cousins the lava lizards can use for shade.

There are also far larger cavities below ground – the remains of tunnels formed during the molten lava's flow. These may now be dry, forming natural traps for unwary giant tortoises, or they may fill up with infiltrated sea and rain water. Such dark, flooded caves are a distinct habitat on Galápagos and they have their own unique fauna (see box above).

It may take thousands of years of weathering and biological action, but eventually the lava surface pales and corrodes while soil accumulates. Volcanic ash or pulverized volcanic cinders (tephra) may help fill crevices, washed or blown from vents above. On a very volcanically active island like Fernandina or Isabela, the

Opposite: Brachycereus *cactus form ancient pioneering colonies on the bare lava. Above: The Galápagos cavefish.*

biological stopwatch on the race to colonize this new land surface may be set back to zero many times as new lava flows over the old. It is a lottery whether the surface will be overwhelmed again, or have time to develop to the mature and stable lowland community that is seen in the arid zone. Eventually all island volcanoes become sufficiently hushed to allow an arid community to thrive. It is in those established pale, dry forests and scrub that the land fauna of Galápagos really comes into its own.

INTO THE ARID ZONE

The first scenery many human arrivals to Galápagos see from their plane or ship is the arid zone. If first impressions counted for much, a lot of them might just go home. Looking at this habitat, it is hard to be very complimentary – it lacks the spectacle of the lava fields or the lush otherworldliness of the higher-altitude habitats. But though the arid zone is rocky, thorny and parched, its occupants are surprisingly varied. On most landmasses, wet habitats like rainforests are more species-rich than dry ones like deserts. Yet paradoxically the opposite is true on these islands – the reason being, of all things, the last ice age.

We usually think of ice ages as phenomena that mostly affected extreme northerly or southerly latitudes. But their impact was global. During the last one (24,000–9000 years ago) water was tied up in the ice caps and as a result conditions on Galápagos were probably far drier than they are today. Evidence from pollen preserved in the ancient sediments of fresh-water lakes such as El Junco on San Cristóbal suggests that at that time the arid zones stretched virtually to the peaks of the islands. With sea levels perhaps some 150 m lower than they are now, the lowland areas would also have been far more extensive. Creatures that survived and adapted to such conditions may have the longest evolutionary history on Galápagos, and are still well represented in the diverse 'desert island' flora and fauna that predominate today.

THE ISLAND GIANTS

Giant-cactus silhouettes virtually define the islands' skyline throughout the arid zone. They appear to be very ancient members of the Galápagos club. The branching trunks of candelabra cactuses stand on the horizon like the lines drawn for an evolutionist's family tree. Reminiscent of the saguaro cactuses of the US, these slow-growing endemics can hold their own in dry forest or on broken lava. It takes an active imagination to see why, but these are the 'dildo trees' described by the first pirates (see Chapter 2).

THE HANDY OPUNTIA

On desert islands it is often claimed that the coconut has all you need to survive. In wilder parts of Galápagos you won't find many coconuts – here, the key survival plant for many species is opuntia. Its cactus pads, though typically spiny, offer vital shade and food for land iguanas and tortoises. With reptilian patience they wait beneath the cactuses for a pad to fall. Though iguanas may rub the pads a little before eating, for the most part they stoically tolerate mouthfuls of spines. In emergencies, people too can squeeze the pads for a little bitter water, and the fruits are edible. The trunks and pads offer nesting sites for many birds, and the flowers are a vital nectar and pollen source too; cactus finches and the rare champion mockingbird appear to be specialists. In an emergency, shipwrecked sailors say the fibrous matter of dead cactus can be strapped to the feet with hide straps as a comfortable shoe sole. What a cactus won't do for you, however, is get you off the island. Its fibrous, hydrostatic trunk is quite unsuitable for making a raft or a boat.

Left: A rare champion mockingbird on an opuntia cactus.

The opuntia cactus tree's sturdy trunks are developed from the sausage-string oval 'leaf' pads that ramify to form branches. It gives them the look of children's drawings and caused Darwin to remark that they were 'great odd-looking' plants. The size and oddness of this cactus varies a little with each main island. That is in part due to the happenstance of their separate history, and in part because different islands offer different conditions to which their inhabitants have become adapted. It is widely believed that they developed trunks to help put their pads out of reach of reptile jaws but, as we shall see, there is a little more to it than that.

Opuntia cactuses have become a linchpin of the arid zone's ecology (see box above). The islands' most famous inhabitants, the giant tortoises, are among those that have come to depend on them there. Tortoises are another member of the Galápagos old school. It is hard not to be moved by a first wild encounter with these legendary creatures, at once so familiar and so strange. There is surprise, often mutual, and then the strong impression of their size, their strength and their tangible antiquity. Even their elephants' feet and oddly individual faces are remarkable. You may well find yourself talking to a giant tortoise – a sincere human compliment.

The arid-zone tortoises have flared 'saddle-back' shells that allow their wearers extra mobility to stretch upwards to forage, and this appears to be a local adaptation. Gigantism is often cited as another adaptation to these arid habitats, but small tortoises survive rather well in dry deserts the world over, and the low-land tortoises are in fact rather less massive than the cool-upland ones we shall meet later.

On Santa Cruz the opuntias grow to be the tallest on any island – so tall, in fact, that it seems improbable anything much smaller than a giraffe could have forced them so high. Rather it seems that competition with other trees for light sent them skyward. It may be that on the more species-rich (and relatively windy) mainland of South America this is a competition no spongy-trunked cactus could have won, but there are far fewer contenders out here on Galápagos.

DRY ZONE FLORA AND FAUNA

Competing with the opuntias in the arid zone is a rather alien-looking plant with long, green, whip branches covered in many hooked spines. Its local name is palo verde and it is just one of the many thorny trees and bushes that make walking here very tough. Another common plant, called chala by the local people, stains shirts and trousers brown with its sap. Consequently an exploratory ramble on Galápagos is often sweet with the turpentine smell of an omnipresent tree whose inoffensive branches are broken to make headway through this otherwise painful habitat. This is the palo santo (the Spanish for holy stick), whose stems are burnt for incense on holy-day processions. It looks dead and grey for most of the year, but primed on the tips of each shoot are buds ready to unfurl dazzling green after the first sufficient rains. This tree survives here by virtue of its patient endurance of drought, followed by a remarkable sprint to make leaves, flowers and fruit. All that growth is often achieved in the time it takes the water from a single large rainstorm to leave the soil. Galápagos trees may compete more below ground for water than above ground for light. In places the skeletal, orchard-spaced palo santo seems to win these hidden contests and often dominates the habitat, ascending the maroon slopes of the large tuff cones almost to the exclusion of everything else.

One of the first creatures to meet you as you enter such areas is the mockingbird. It hops and runs as much as it flies, covering short distances with low, swooping glides. On some islands you may soon have a small family of them around you. They personify the islanders' naïve trust of newcomers, taking a close interest in anything new that enters their territory. These birds are very aware that with change comes opportunity, but risk doesn't seem to enter the equation. On Española mockingbirds may be into your drinking water before you are. On Fernandina serious fights between land iguanas even attract mockingbirds looking for spilt blood. But on Floreana the birds paid the ultimate price for their fearlessness – decimated by introduced cats and now confined to Champion and Gardner islets, where the cactuses in which they nest and feed still thrive.

Darwin's finches are here too, in a profusion of little grey and black blurs. They too are marking time until the next rains bring the flush of food they need to breed. At first sight they could be mistaken for being all of a kind, with dark males and brownish females. But on looking more closely you can see there are different forms and sizes among them – particularly their beaks. While they are famous for their nondescriptness, their behaviour is anything but ordinary. Some of the small-beaked finches found on Wolf have addressed their island's water problem by eating eggs and becoming vampires, breaking eggs by rolling them over cliffs and pecking blood from wounds they inflict or worsen on seabirds. It is a habit probably acquired during their more benign routine of picking-off ectoparasites in nesting colonies.

One finch you won't confuse with the rest is the warbler finch, which has a decidedly unfinch-like small, probing beak and a melodious song. With a paucity of true warblers, it has virtually become one, filling a vacant niche as finches here are

THE LOST FINCH TRIBE OF COCOS

Any guide will tell you Darwin's finches are an endemic group unique to the Galápagos. Well not quite, there's one that got away. On the tiny, lushly tropical island of Cocos, 500 km from Costa Rica, lives a very lost Darwin's finch. What storm or hopeless migration brought it there 700 km from Galápagos is anyone's guess. What is more interesting is that there is just one species of finch on Cocos, and one it is likely to stay.

Cocos is a solitary speck of an island, rather than being part of a larger archipelago. It therefore offers no opportunity for the sort of population separation that helps get species splitting started and led to 14 species of finch evolving from just one colonizer on Galápagos. So instead of radiating as different specialist species, as Galápagos finches have done, the Darwin's finches of Cocos become specialists as individual birds. Hence some choose to major in eating insects, others flowers and yet others fruit and seeds. It appears the birds pick a trade that will put them beyond the skills of other neighbourhood finches, so that even without speciation an individual can forage with less competition from its fellows.

Previous page: A Galápagos hawk watches over the caldera of Alcedo volcano from the sulphurous fumaroles on its northern rim.

prone to do. Another large finch found in the trees eats leaves and buds, something very few other birds can do – the vegetarian finch. Most surprising of all is a somewhat hyperactive finch sometimes encountered with a carefully prepared twig held in its beak. The woodpecker finch is normally seen on dead branches, intent on any holes in the decayed wood. When things look promising it draws back its head and plunges the spine into the hole. Deftly manipulating this tool it will carefully draw out the head or tail of a large larval insect, eaten with a well-timed lunge. Only a true woodpecker or the aye-aye lemur could have done it better – but they are not here.

It is often said that these creatures are so tame because there are no predators on Galápagos. From a human perspective that's true, but if you are the size of a lizard then the snakes that patrol the arid regions of all but the northerly islands are formidable adversaries. There are four currently recognized species of constricting snakes, and the best way to find them inland is to look for the noisy gangs of finches or mockingbirds that mob them on the ground. Snakes and perching birds have their own mutual predator in the form of the Galápagos hawk, a handsome bird that sits resolutely atop the islands' food pyramid. In a celebrated incident Darwin found he could push a wild hawk off a tree with the barrel of his gun. Despite such past breaches of trust, the hawk seems just as tame today.

THE TRANSITION ZONE

Gaining altitude, starting between 80 and 120 metres, the habitat changes. This is the transitional zone. The higher you climb here, the more rain falls. It is still not much. Many of the trees and bushes are in leaf, but the leaves are waxy (e.g. guayabillo) or furry (e.g. pega pega) to help hold on to moisture, and the roots are deep (sometimes 10 m or more). Some plants, such as palo santo, are familiar from lower down the slopes, but they are less dominant here. There is no closed canopy of trees, so a diverse array of woody shrubs can thrive. Crawling and flying among them is a rather scarce invertebrate fauna of crickets, colourful painted locusts, butterflies, beetles, ants and spiders, including an endemic black widow. Having seen only a handful all day, you may start to wonder where the other couple of thousand species are. The answer is that most lie low while it is dry, others are nocturnal and many are simply tiny, rare or belong to another island.

NATURAL PARTNERS

One strange feature of the arid and transition zones is that most of the endemic plants you meet have bright yellow flowers – Darwin's cotton, spined leocarpus, Galápagos purslane and palo verde to name just a few. It may be more than a coincidence. An

THE VOLCANIC CALDERAS

As we saw in CHAPTER 1, on the younger islands the craters of giant volcanoes collapse, as their magma chambers are emptied at depth, to form vast central calderas. Several habitats intermingle on the rims and caldera floors, but are a unique environment by virtue of their active geology and unusual topography. On Fernandina the caldera walls are so steep that they host almost daily landslides, sometimes involving hundreds of tonnes of material. Vegetation finds few root-holds around these unstable cliffs and terraces. That barren instability does not stop female land iguanas making a perilous descent into the caldera to lay their eggs near the warmth and moisture of the active fumaroles. It's hard to picture a more dragon-like thing to do. The floor of Fernandina's caldera includes a tranquil-looking mineral lake several square kilometres in area, which in years gone by has harboured a population of pintail ducks. The lake is anything but tranquil, however – in the last 50 years it has boiled away several times when entered by tongues of molten lava and it has been displaced from its lakebed and slopped up the wall of the crater as a result of massive landslides. It has even been dropped hundreds of metres to disappear for some years into the bowels of the Earth. By all accounts the ducks went with it each time.

The caldera of Alcedo volcano on Isabela seems a far more peaceful place. Dome-backed giant tortoises roam in their hundreds, creating Elysian fields of immaculately rolled and cropped pasture interspersed with picture-perfect bathing pools. But the active fumaroles, deep pumice flows and shattered obsidian blocks hint that this volcano too has a darker side. Genetic evidence from the tortoises suggests that about 100,000 years ago Alcedo erupted so

Left: A view into the remote caldera of Cerro Azul.

devastatingly that it wiped out all but a handful of them. More recently Galápagos naturalists have found tortoises with rocks embedded in their shells after smaller explosions near the fumaroles.

It used to be thought that Alcedo was the last great bastion of the giant tortoises, but during goat eradication on neighbouring Wolf volcano in 2004 biologists were amazed to find tortoises in their thousands present on its flanks. Even land iguanas were abundant right up to the 1700-m rim. More remarkable still was the corroboration of a scattered population of pink land iguanas, which, at the time of going to press, is unidentified as to being a new species, race or perhaps just mutation.

It is hard to visit the calderas of Fernandina, Alcedo or Wolf, but the largest Galápagos caldera of Sierra Negra on Isabela is easily reached from the town of Puerto Villamil. Its caldera floor was, until recently, an example of an arid zone developing on an old lava flow. Events of 2005 changed all that as, almost overnight, hundreds of thousands of tonnes of molten lava flowed in to fill over 10 sq km of the eastern caldera with red-hot lava. At the time of writing it is a smouldering basin of hot rock quite devoid of any life – but of course it won't be long before the lichens and spiders arrive.

Above: Humid epiphyte-clad forests soften the higher volcanic slopes of the older islands.

oddity of the islands' arrival lottery is that only one really dedicated daytime pollinator ever made it to Galápagos, and now its services are greatly in demand. The carpenter bee's favourite colour seems to be yellow, and the massive black female speeds with an alarmingly deep buzz from flower to flower, tirelessly provisioning for her young. Wherever this bee is absent, as it is on the northern islands of Genovesa and Marchena, birds such as Galápagos doves step in to pollinate flowers. Opuntia cactuses on these islands accommodate these feathered partners with unusually soft spines.

The other main hue of endemic Galápagos flowers is pale cream to white, colours often worn by tube-shape flowers. Those structures and hues are reserved principally for the hawk moths that dominate the pollinators' night shift. They can often be seen hovering around suitable flowers during an evening out in the town of Puerto Ayora on Santa Cruz. A prime example is the endemic lava convolvulus, which has trumpet flowers with tubes 10 cm long – the same length as the green hawk moth's extraordinary proboscis. This is about as choosy as a Galápagos plant ever gets – it's a risky business being too exclusive with so few pollinators around.

Above: The scalesia zone near Los Gemelos pit crater on Santa Cruz.

CREATURES OF THE NIGHT

Night is also a common time to meet Galápagos's most feared land inhabitant – the scolopendra centipede. Growing up to 30 cm long and with a habit of turning up where people least want it, this phobia-inducing predator is capable of tackling creatures the size of lava lizards; they have even been observed killing juvenile rats. Its toxic bite is a powerful weapon – though not sufficient to deter the hawks and herons (and, more recently, town cats and chickens) that commonly feed on it.

As well as invertebrates, the night shift in the arid zone also reveals geckos and the islands' only naturally occurring terrestrial mammals, bats and rice rats. The bats almost certainly flew here under their own power, and the rice rat's ancestors must have made the journey by rafting. The rice rats thus have the distinction of having made the longest unaided oceanic crossing of any mammal on the planet. Little is known about either of these native groups, though neither is particularly shy. The rice rat will tentatively nibble the toes of researchers sleeping rough on Fernandina, and the bats can often be seen attending certain favourite streetlights in towns. Barn owls hunt here too, and it may come as a surprise that these birds are of the same species that you have probably seen back home (they are found on every continent except Antarctica). The barn owls' habit of taking prey back to their nests in lava tunnels led to the preservation and discovery of the sub-fossil bones of a giant rice rat (extinct) and an otherwise unknown species, christened Darwin's rice rat, on Fernandina in the 1970s. That species was found alive in 1995, followed a few years later by the rediscovery of yet another rice rat, previously presumed extinct, on Santiago. It leaves four species of rice rat alive today. Three others are recently extinct, having lost out to the introduced black rats. The Santiago rice rats probably only survived because they can feed on watery opuntia cactus – a native skill black rats just don't have.

Above: Owls, like this short-eared owl, carry their prey into caves that preserve the sub-fossils of past faunas.

On many of the islands that boat tours visit, the transitional zone is the end of the story. It marks the peak altitude of the outlying older parts of the archipelago and hence defines them as true desert islands. But on the larger islands of Isabela, Santa Cruz, Santiago, San Cristóbal and Floreana, you can keep climbing. Not all visitors to the islands get there, but perhaps they should.

THE MOIST UPLANDS: THE FORESTS OF THE SCALESIA ZONE

From about 300 m above sea level on the south side of Santa Cruz (several hundred metres higher on the north), a small change in height brings a remarkable change in habitat. As you climb, the transitional zone slowly becomes lusher and more richly adorned with epiphytes and herbaceous plants. You have left the arid lowlands behind you and are entering the first of the moist upland habitats.

In the garúa season, the cool sea air condenses heavy mists at these altitudes and the scenery is often transformed into a drifting shadow play of gnarled silhouettes. The habitat eventually becomes a crowded umbrella canopy of intense green. From a distance it could be mistaken for a plantation, but it is in fact a natural Galápagos forest of the endemic sunflower tree scalesia.

Viewed from below, the canopy of the scalesia forest is a thin mosaic of lance-shaped leaves. It admits enough light for a densely interwoven community of epiphytic 'air plants' and ground plants to thrive beneath. Ferns, orchids, bromeliads, grasses and mosses are abundant here, though not particularly diverse. They are latecomers in terms of colonization, with tiny seeds that may

originally have blown to the islands from the mainland, or arrived in the ruffled feathers of birds. Unless they will also grow on lava, they cannot establish before the trees they grow on are here. In fact the number of endemic species is surprisingly low, a result perhaps, of this habitat only appearing relatively recently after the ice age (see Into the Arid Zone, page 92).

Above: Vermilion flycatchers provide dashes of brilliant colour in the moist-zone forests.

At its best the scalesia forest looks like a good setting for a Grimms' fairytale. Warbler finches sing among a twisted, low canopy of leafy branches. Vivid vermilion flycatchers zip through the soft greenery, past the predatory yellow gaze of short-eared owls. This too is the haunt of Methuselahan dome-backed tortoises that in their dotage carry a world on their backs in the form of tiny lichens, algae, invertebrates and mosses. The tortoises' traditional tracks criss-cross these higher transitional and scalesia forests on black mud trails that lead eventually either to water or to the nesting sites in the transition zone below. Dome-shaped shells help during their tank-like negotiations through dense vegetation while not impeding their grazing on low herbs and grasses.

The tortoises' exceptional size and habit of sleeping in excavated soil bunkers insulate them from the cool night air at these altitudes. Size also helps

THE DEAD SNAIL TRAIL

The observant on Santa Cruz may notice frail white snail shells crumbling to powder among rocks and crevices. Their remains are scattered from high to low in the islands. They range in shape from long and thin to rounded and robust. When alive, their shells were dark, striped or mottled, but you probably won't find a live one to check.

You are witnessing the mysterious and tragic extinction of not one, but many species of endemic land snail (scientifically *Bulimulids*) that were once very abundant here. It is estimated that there were upwards of 25 species on Santa Cruz and over 66 in the archipelago – an evolutionary radiation that greatly exceeds that of the more famous finches. Many were never described. It seems patches of forest or valleys isolated by lava flows were capable of producing their own snail types, much as whole islands do with mockingbirds.

The reason for their rapid and ongoing extinction at the end of the twentieth century is not known. Humans are implicated, but other species of this land snail are still quite abundant on Floreana, where there has been long-term human impact.

Certain species are still reasonably abundant on the rims of outlying Galápagos volcanoes. But with so much else to conserve, they may slip into oblivion without anyone ever noticing when, or understanding why.

decide contests between tortoises. Fights begin with aggressive neck stretches that allow competitors to size each other up and, if neither backs down, develop into a bout of serious leg nipping. Perhaps because males fight more, they are larger than females. Evolution seems to work that way.

THE BROWN ZONE

Above the scalesia zone, wherever it has not been cleared for farming, is a habitat often called the brown zone because a profusion of lichens drapes the dominant cat's claw trees and makes them look brown during the dry season. Anyone who has been repeatedly snagged by cat's claw might even gain a little satisfaction from seeing the great burden of epiphytes its recurved thorns cause it to suffer here. In fact, cat's claw is not limited to the brown zone. Its spines represent a near impenetrable barrier on many islands. On the southern crater rim of Alcedo, cat's claw, dense with mosses, once interlocked with a wonderland of tree ferns, herbs and grasses to create a 'green zone'. It caught water from the misty clouds and dripped it into extensive pools where massed tortoises drank and bathed. Goats, fairly described as the Genghis Khans of nature conservation, have lain waste this habitat. Few trees and pools are left, but with the goats now effectively being eradicated, there is real hope that the vegetation will one day recover.

THE MICONIA ZONE

On Santa Cruz and San Cristóbal, the scalesia forest is not the only moist habitat virtually dominated by a single species. At a height of about 600 m on the southerly face of Santa Cruz, another bushier broad-leaf plant carpets the higher slopes of the ancient volcanoes. It is the Galápagos miconia, an endemic with patterned leaves and small purple flowers. Up here it is interspersed with extensive bracken stands, grasses and a few flowering plants such as ground orchids and St John's wort. Walking through this vegetation you may well spy a vivid red eye skulking among the ground-litter and if you wait quietly the exquisite little Galápagos rail may approach to see what has disturbed its patch. The miconia zone represents a last stronghold for this species, which has been wiped out by marauding rats and cats lower down the slopes. The rail is vulnerable partly because it is nearly flightless – a common feature of island evolution among birds and insects. The explanation often lies in a native lack of predators, compounded by the simple fact that those individuals that fly away from an island may never return to breed. In fact it is the least mobile creatures that often develop the most diverse island families – as the box on page 82 explains.

HIGHLAND BOGS: THE FERN SEDGE ZONE

You are nearly at the top now. Above the scalesia and miconia zones, on the more exposed summits of older islands, is the fern sedge zone. Resembling moorland in its luxuriance of mosses, sedge and grasses, this zone is generally rather devoid of animal

FARMLAND AND INTRODUCED SPECIES

Though the Galápagos National Park covers 97 per cent of the land surface of these islands, a disproportionate amount of the limited moist zone falls in the remaining 3 per cent. As a consequence a great deal of it has been cleared or altered for farmland – and what remains has been badly affected by introduced plants and animals. The miconia zone (see above) has been taken over by quinine trees in many parts of Santa Cruz, and mora, a relative of the raspberry, is spreading across wild habitats and farmland alike with its dense mats of spiny stems. On the Sierra Negra volcano on Isabela the introduced guava bush all but dominates the volcano rim's flora, its seeds spread in the dung of feral domestic animals – much as tortoises once spread the seeds of the endemic tomato. Indeed new pest relationships seem to be replacing the old all over the archipelago. Fire ants, the scourge of the endemic insects, appear to tend and to carry introduced scaly bugs that reaped similar havoc on certain endemic plants (until a non-native ladybird was introduced to control them). It will be a constant war to keep on top of these pests and in many places the current battle is lost. It is a sad fact that Galápagos now has nearly 20 per cent introduced vertebrates and insects, and will soon have more introduced plants than native, changing its ecology for ever.

Right: The highlands of Santa Cruz – miconia and bracken stand in surprisingly lush contrast to the coasts beyond. Opposite: Colourful bogs with floating red azolla, fern and sedge are found at the very summits of the older islands.

life. The rainfall here can average 1.5 m per year, but can be 3 m or more in a wet year, an amount of water that any rainforest would be happy to receive. The dense accumulations of organic compost seal the porous rock surface, forming a soft and boggy mat that in places settles into deep peat beds. There is standing water here, often covered with red and green azolla fern mats. The tallest local plant, standing head and trunk above the rest, is the sporadic Galápagos tree fern, which reaches 3 m. It is a plant little changed since the Earth's first primeval forests grew 400 million years ago. It somehow seems appropriate that it stands silent watch from these heights.

As the hawk flies you have travelled only 10 km or so from the ocean. Yet from such summits, green and drenched in water, the parched coasts seem a world away.

PETRELS: SYNCHRONIZED FLIERS

At certain times of the year, on some of the high peaks of Santa Cruz, San Cristóbal, Floreana and Santiago islands, a hauntingly strange call at dusk is the first inkling you have of something remarkable. Wait, and soon you will be surrounded by the sound of elegant backswept wings cutting the air at high velocity. The Galápagos petrels have arrived. These endangered birds, relatives of albatrosses rather than the smaller storm petrels, with whom they share only a name, nest in burrows in the highlands. It is remarkable how they can find their homes in the misty darkness, but they do – and their single chick is fed on the part-digested fish and squid they bring back from the ocean. As night progresses the unearthly din of their cries rises and falls as the birds call to one another during synchronized aerial acrobatics at remarkable speeds. It is one of the least known yet most spellbinding natural spectacles on the islands.

5

Stretches of rugged lava, battered cliffs, mangroves and idyllic beaches create a *dynamic and diverse coast*, and a concentration of Galápagos life. Though crammed into a narrow band running the length of the shores, the creatures of the coast are some of the best known and *most fascinating* in the islands. Even more than the land dwellers seen in CHAPTER 4, these animals are influenced by the cycles and seasons of the sea, and their lives intertwined with the *productivity* or *fickleness of the waters* around the archipelago.

SURVIVAL ON THE COAST

Previous page: A brown pelican scans the wave-battered shore for a suitable fishing site; the coastline of Galápagos can be both productive and perilous. Above: A Galápagos sea lion rests alongside flamingos in the lagoon on Rábida Island; such scenes illustrate the unique combination of creatures, often from very different regions, encountered on these coasts.

BETWEEN THE DEVIL AND THE DEEP BLUE SEA

Almost everywhere in Galápagos is infused with a sense of the sea. That these are true oceanic islands, never connected to the mainland, is a hard fact to forget. Seen from the deck of a boat, it is the vast Pacific that appears constant, interrupted by arches of islands that defiantly break its surface. Once on land it is nearly impossible to leave the sea behind and out of mind. In Galápagos, the coast has been a draw for life since the first volcanoes raised their domed peaks above the surface of the sea and the first creatures swam, flew or floated ashore.

But why is so much of Galápagos life concentrated on her shores? And what is the creative force remarked upon by Darwin that has led to this life becoming such an assortment of creatures? The coastal fringe is the margin where the lava islands, born from the hellish depths of the Earth, meet the rich mix of oceanic currents and upwelling abyssal blue sea. It's a combination that in places makes for a staggeringly productive coast. The 13 main islands and over 100 rocks and islets have a collective coastline that stretches for 1350 km and provides a myriad of niches for life to fill.

Coastal life is characterized by hardy plants that can tolerate arid salty conditions and by those animals that use the land as their springboard to gain access to the resources of the sea. But despite the seeming profusion of life, this is a hard place in which to survive. The challenge to cope with the lack of fresh water, and at times searing heat, while carving out a living in fickle, cool waters is a tough one. In their dependency these creatures are at the mercy of the sea and are greatly affected by the El Niño phenomenon and an increasingly heavy human hand.

The coasts of Galápagos offer wildlife watching on a plate. They are the focus for most boat-based tours that daily hop from island to island and go ashore from site to site. And rightly so. A trip here is like watching a wildlife documentary, offering a front-row seat of uninterrupted behaviour with all the sights, sounds and smells of real life. Binoculars are not a requirement! It is an immediately satisfying and gratifying experience in a way many of us have grown accustomed to expect in the convenience of modern life. To the better informed, however, in the scenes playing out before our eyes can be glimpsed evidence of the processes and forces at work, displayed with a fascinating clarity particular to these islands.

Reviewing the facts here given, one is astonished at the amount of creative force … displayed on these small, barren and rocky islands.

Charles Darwin,
Journal of Researches (1845)

But what are these forces? The oceanic isolation and climatic conditions are the principal physical factors that lead life to drift genetically, to adapt and take on unique forms – traits that characterize much of the life of Galápagos and for which it is famed. It is these 'creative forces' that lead to the fascinating mix of endemics, world travellers and local residents that make up the coastal life of the islands. The best way to meet this unique collection of creatures and explore how life has made a home on Galápagos's shores is to journey through the habitats of the coast, across the lava shores and through the mangrove glades, to the sandy beaches and salt-water lagoons, and finally to the cliff tops of the oldest islands.

LIFE ON THE ROCKS: THE SUN WORSHIPPERS

Rough lava shores dominate the coastline of Galápagos. The rocky coast is a forbidding mix of aa and pahoehoe lava and wave-washed rock, in places eroded by the sea to boulders and subsided slopes. It receives the brunt of the ocean's energy and sits in the full strength of the equatorial sun.

Thermal regulation is key to surviving on the rocky shore and the marine iguana has evolved like no other creature to master this. In doing so it has developed

A ROCKY FOOTING

Right: Marine iguanas crowd the best basking spots on Fernandina. This behaviour, developed in part due to the absence of terrestrial predators, also made the marine iguanas easy pickings for early mariners in need of fresh meat.

The marine iguana has adapted not just to survive but to thrive on Galápagos's coasts, surrounded as they are by cool, productive waters and characterized by warm, sun-baked lava shorelines and a lack of terrestrial predators. In Cape Douglas, on the shore of Fernandina, thousands of marine iguanas can be witnessed undertaking the morning warm-up routine. Across the archipelago over 200,000 start the day in this way. It is a sight to behold and a ritual that has played out most likely every day for millions of years – but not always on these islands! For the marine iguana is a species older in existence than the very rock upon which it basks. Genetic research indicates that a divergence between Galápagos land and marine iguanas occurred approximately 10 million years ago and suggests that it took place in the islands and not on the mainland. This puts the arrival of the ancestral iguana species on Galápagos – floating in from mainland South America on rafts of vegetation – before that time and long before the emergence of the current islands. The aquatic ability that then further developed in the marine iguana has not only allowed the single species to spread through the islands but also ensures it is an island hopper, remaining as a single species across the archipelago and able if necessary to move to new islands on the tectonic conveyor belt of Galápagos island formation. It is truly one of Galápagos's oldest endemics.

A LIZARD'S LEGACY

Lava lizards can be seen across the islands, with the exceptions of Wolf, Darwin and Genovesa. But unlike the marine iguana they are not a single species. Seven species have evolved on seven different, central islands, providing a fabulous example of adaptive radiation (see page 87). The different species come in a variety of colours and sizes depending on which island you are on, with males typically being more colourful, patterned and larger than the red-throated females. And within species there are differences in appearance. Populations living near the beaches, for instance, have become greyish-yellow, while those among the lava rock have taken on a darker hue. Most fascinating, however, is the fact that the head-bobbing 'push-up' routine used to defend territories is not only specific to each species but to populations of the same species. It's like a hip-hop dance-off in which the prize is the best street on the block! Research into the extent of genetic variation among the species suggests that they may have evolved from multiple colonial ancestors, but the jury is still out on this. Regardless, the genetic differences that have developed are, to a large extent, due to such strongly territorial behaviour, which separates populations and ultimately leads to significant separation occurring at species level.

a unique relationship with both sun and sea. These spiny, tough-looking creatures best embody the bleakness and ruggedness of the lava coast. As the first warming rays of the morning hit the dark lava, the marine iguana colonies stir. It is as if the very coast itself crawls. Iguanas move into position and begin to bask. Their body temperature can rise up to 40°C and it is this heat that will allow them to enter the cool waters to feed. In the early morning they lie prostrate and in profile to the rays, maximizing the area of dark-coloured body mass exposed to the warmth. Reptiles living in the tropics need relatively little energy intake to survive, as the high climatic temperatures allow an optimal body temperature to be reliably maintained. A 1-kg marine iguana can get by on 37 g of food a day, a tiny amount for such a strict vegetarian.

Another coastal reptilian resident also stirs in the first heat brought by the sun. Lava lizards emerge from their night-time nests to bask in the early morning rays. Nesting in the crevices and crannies of the lava conserves heat during the night but, like the marine iguanas, these lizards must top up their temperature at the start of the day. That done, they spring into life to feed and defend their territories and maybe find a mate in a most ardent display of head-bobbing (see box above). They are omnivorous but can most easily be seen on prominent rocks and prime positions within their territories, catching insects that skim over the undulating lava landscape.

As the sun climbs in the sky, marine iguanas head towards the sea. Though their reliance on the warming rays of the sun is apparent, recent research shows that

it is actually the influence of the lunar cycle that triggers the iguanas to move. Why in Galápagos did this otherwise land-based creature develop such a link to the ocean and a propensity for entering the water? The reason can be seen in the dry, scrub-lined coast that offers little in the way of sustenance. A falling tide peels back the sea to reveal expanses of short green and red macrophyte marine algae covering the rocks, and it is on this that the marine iguanas begin to feed. They appear uniquely designed for such a specialist lifestyle, reliant on this single resource. Powerful limbs enable them to grip wave-washed lava rock and flat-fluked tails propel them in open water, diving for minutes at a time. Though apparently sparse, these algal carpets are a deceivingly productive resource. Darwin himself puzzled over the sustenance of these 'imps of darkness', for at first glance the cover of algae appears minimal and hardly likely to support such populations. But the combination of cool, nutrient-laden waters and the strong equatorial sunlight means algae grows rapidly – it is able to double its length in a fortnight. Indeed the tenfold difference in size among populations of marine iguanas spread throughout the islands illustrates how varied algal productivity is across the archipelago. The smallest iguanas are found in Genovesa, located in the northeast, which is most influenced by the warm, inter-tropical convergence zone that supports less algal growth, while the largest are found on the west coasts of Isabela, washed by cool, productive, upwelled waters.

Above: The marine iguana is the only marine-going lizard in the world. While small individuals can survive on the pickings in the intertidal zone, larger animals must go further offshore and dive deeper to find sufficient algae.

As marine iguanas rest, ever opportunistic sally lightfoot crabs pick dead skin and parasites from between their scales in an act of mutualism that benefits both. The adult crabs stand out as vibrant punches of colour, while dark juveniles blend into the lava rock. There are over a hundred species of crab in Galápagos and its surrounding waters, but sally lightfoots are the most apparent. They are the

'THE INCREDIBLE SHRINKING IGUANA'

As if marine iguanas weren't already fascinating enough, research into their response to food shortages associated with El Niño events suggests that they even have the ability to shrink during lean times. They are believed to be the only known adult vertebrate able to do so. Studies show that some individuals shrank in length by up to 20 per cent over two years, an observation that was particularly evident in larger individuals and comparatively greater in females than males of the same size. It's too great a change to be explained by decreases in cartilage and connective tissues, which suggests that bone absorption causes much of the shrinkage. Regrowth follows when conditions improve and indicates a further adaptation to a relatively frequent climatic variation.

unequivocal shore cleaners, present on all rocky coasts across the archipelago, feeding on seaweed and organic material found on shore and at times on each other. Other forms of mutualistic cleaning in the coastal zone can be seen in tidal pools, such as those on Española, where marine iguanas float on the surface, posturing with legs splayed, to invite rainbow wrasse to pick their skin clean.

In the height of the hot season between February and May, temperatures on the lava rock can reach 50°C during the middle of the day. Overheating is a strong possibility for marine iguanas basking on their return from feeding. Once sufficiently warmed up, like Incan deities they orientate to face the sun, minimizing exposure to its rays. They also raise their heads and chests to allow air to circulate beneath, or seek shade under rocks and shrubs. The sight and sound of snorted salt excretions – extracted as a by-product of their algal diet by their highly efficient salt glands – can be seen and heard across the colony. In the constant battle of thermal regulation, lava lizards take to the 'highlands' to find any hint of a cooling breeze. These 'highlands' could be any prominent rock, snoozing sea lion or the head of a marine iguana within the lizard's territory. Standing on stretched limbs they too raise their bodies to allow the air to pass under their bellies. Later, marine iguanas will cluster together to conserve heat collectively through the night while some lie under vegetation to do the same. After a frantic day on the rocks, lava lizards will take to their nests to rest for the night.

The marine iguana's reliance on a single food source can spell disaster. El Niño years bring sustained pockets of warmer waters that lead to green and red algae species being replaced by brown algae, which marine iguanas find hard to digest. In the 1982–3 El Niño the islands' population crashed by 70 per cent, and in 1997–8 by 50 per cent. They bounce back in the La Niña years that typically follow, though, and there is evidence that some individuals pull through by literally tightening their belts (see box, page 115). But human activity also takes its toll. Evidence following the oil spill after the tanker *Jessica* grounded in 2000 shows it was the marine iguanas that were most affected by the spilt oils and dispersants used in the clean-up operation. The population on Santa Fe suffered a 62 per cent mortality rate in the following year. Ingestion of the foreign chemicals is thought to have affected the digestive bacteria that enable the iguanas to break down the algal cell walls. It's a telling example of the fragility of such a finely evolved lifestyle.

FEELING THE HEAT

In the struggle for temperature control in the heat of the rocky shore, the challenges faced by the Galápagos penguin couldn't be more different than those of the reptiles. But how does a species originating from the sub-arctic manage to live and breed successfully here in the tropics? The penguin's seccret to survival is their continual reliance on the cold waters of the Humboldt current that first brought their ancestors here from the southern coasts of Peru. On the lava coast of Fernandina, Isabela and Bartolomé they are often in the water before sunrise. For the penguins the heat of the day is best avoided by taking to the cool waters of the Bolivar Channel and the other tongues of cool water that sweep around sections of the western and central islands. After carefully preening their insulating plumage in the shallows, they head out for a day at sea. As you will see in CHAPTER 6, it is the cool Humboldt current and the Cromwell upwelling that are the driving forces of productivity for the archipelago and support the shoals of mullet and sardines upon which the penguins feed.

Later in the day the braying calls of a penguin colony echo along the coast at Punta Cristobal, Isabela, while with a typically comical clamber returning residents can be seen leaving the sea and waddling to their nesting grounds. Their head-swaying walk is actually a means of communicating to neighbours that they are just passing through another's territory. The fickle productivity of the waters around Galápagos is clearly reflected in the breeding strategy of the penguin. A drop in sea temperature below 24°C stimulates the onset of breeding and if conditions are favourable the penguins may have up to three clutches a year. To further avoid the heat they take advantage of the natural protection and shade given by the lava itself, nesting and retreating into lava cracks and crevices. All the walking and waddling is hot work. To cool themselves actively they shade their feet, which act like radiators, and further dissipate heat by flapping wings to expose skin underarms rich in blood vessels to the air. Panting rapidly, a process termed gular fluttering, also helps to cool the penguins as they battle against the equatorial heat.

Opposite: Marine iguanas share the rocks with neighbourly sally lightfoot crabs. The crabs are so named for their ability to skip short distances across the surface of the water in rock pools and gullies along the lava coast.

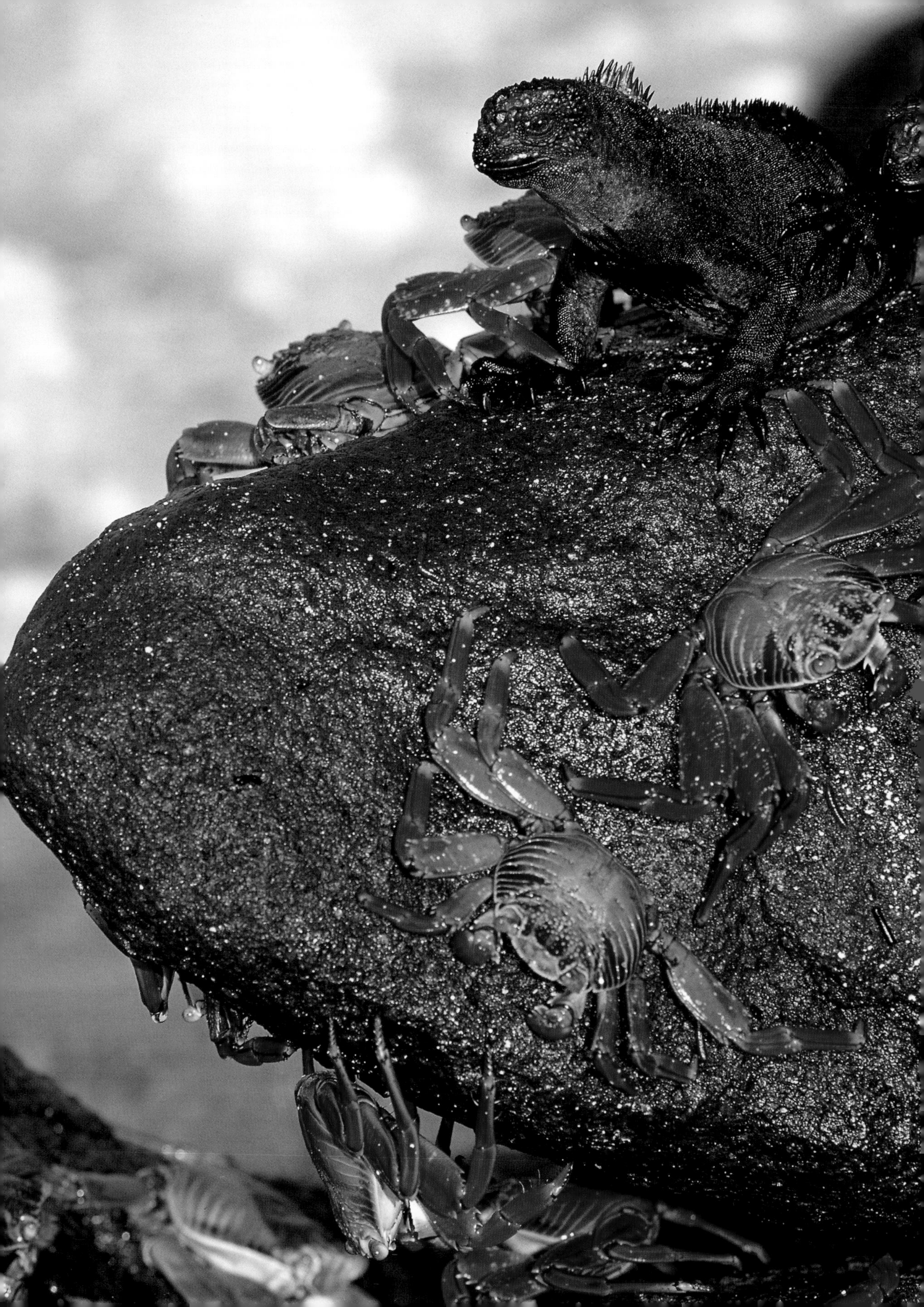

COASTAL RETREATS

Above: A flightless cormorant hangs what wings it does have out to dry. Though its flight feathers are now of no use, long tail feathers help it to steer while diving, and a covering of dense plumage provides insulation against the cool waters encountered around the western islands.

The lava coast provides shelter for another species with southern origins. Galápagos fur seals spend the day mostly avoiding the heat by sheltering in the shade of lava ledges and in underwater grottoes, formed by networks of broken lava tubes and flows in islands such as Santiago and Marchena. Snores and snorts of territorial squabbles mix with the reverberations of water moving through the grottoes. Despite their name, Galápagos fur seals are actually a type of sea lion, with their closest relative, the southern fur seal, being found around the southern tip of South America and up as far as southern Peru. The Galápagos fur seals are the only tropical species of an otherwise subantarctic genus. Their thick pelt, evidence of their southern origins, is ill suited to the tropical sun, but serves well to protect them against the cool water, especially given their smaller body mass, which loses heat more rapidly than that of their larger sea lion cousins.

In the evening light the fur seals can be seen preparing to head out to sea to feed. Their nocturnal activity illustrates an interesting adaptation to Galápagos's coastal conditions. Bulging large eyes, like comical glasses, enable them to hunt in the crepuscular period and throughout the night when squid and other plankton

PRE-PREPARED PENGUINS

Left: When conditions are good and fish abundant, Galápagos penguins hunt successfully in groups. When sea temperatures warm up and fish become scarcer the penguins forage in very small groups or alone.

In a place where the principles of adaptation and evolution are commonly exemplified, the ancestors of the Galápagos penguin arrived at the islands already equipped with their fair share of adaptations to a life in waters like these. Wings adapted to fly underwater and a torpedo-like profile make these birds immensely effective at finding and feeding on the patchily distributed schools of fish. The polarized markings protect against predation and aid in stealthy hunting. Light bellies blend better with the sunlit surface waters when seen from beneath; from above, black backs stand out less against the deeper, darker water. Such markings also potentially help to disguise against more recent threats; penguins have been observed turning their backs to feral cats, perhaps in an attempt to blend into the dark lava rock.

migrate vertically to shallower water. Research into the fur seals' diving strategies reveals that they are essentially surface feeders and rarely dive below 30 m – in stark contrast to Galápagos sea lions, which regularly feed down to 40 m and can top 100 m – and so are only able to tap into the plankton resource while it ascends during the night. They may therefore also avoid hunting during full moon, when the extent of vertical migration is lessened and when they also come under greater risk of attack by sharks.

WATER WINGS

Under water, much of the rocky shore consists of a continuation of the rough lava coast and boulder fields, encrusted with barnacles, sponges and anemones, or carpeted with algae. Submerged marine iguanas can be seen clinging to rocks in the surge, cropping the neat fields of algae with the serrated teeth that line their mouths. They are not in a rush. Seaweed isn't going to escape – unlike the food of the endemic

Above: The spectacular sight of plunge-diving blue-footed boobies can occasionally be seen from shore. The males are lighter in weight than the females, with longer tail feathers, giving them greater agility in the water and making taking off easier.

flightless cormorant. Diving from the surface, the cormorant must swim down against the buoyancy of its plumage, leaving a trail of small bubbles in its wake. It erratically criss-crosses the boulder fields and bobs up to the surface. Once again it dives. Large, webbed feet propel it down and across the boulders and lava, its long, hooked beak probing cracks and crevices for fish, eels and octopus. Its technique of surprise and agility is an effective one. After a number of successes it returns towards shore, picking up a small branch of seaweed before it leaves the water. On land, stubby wings are hung out to dry in a seemingly redundant behavioural throwback to its evolutionary past. Its turquoise eyes bring a striking beauty to an otherwise comical-looking scene. Arriving back at the nest, in a show of ritual bonding, the returning bird presents the seaweed as a gift to its partner, who adds it to the nest and hands over custody.

The flightless cormorant is a true oddity and one of the rarest species of bird in the world, developed in an evolutionary cul-de-sac to live only on the shores of western Isabela and Fernandina. You couldn't wish for a clearer example of the theory of adaptation and evolution. Their scraggly wings give the impression that you've come across them halfway through the evolutionary process, caught semi-naked in the act of changing. In the absence of terrestrial predators and in the presence of a sufficiently reliable concentration of food, the need for flight was lessened, while the need for greater agility and dynamism during fishing remained. Less obtrusive wings emerged as the most efficient form.

The phenomenon of flightlessness isn't that common in seabirds. Flight is a powerful aid to colonization and most seabirds are well adapted to covering large distances; indeed, they were probably the first creatures to take up residence on new Galápagos islands. For most the 1000 km from the mainland to Galápagos isn't too far and the rich waters and original absence of land predators offer a warm welcome. The high rate of endemism in Galápagos land birds (79 per cent) is in stark contrast to that of seabirds, with only 5 of the 17 species found only on Galápagos. While the upwelling waters around the archipelago offer a rich fishing ground, it is the rocky coasts that provide the space these birds need for creating colonies, their one link to the land.

During the height of the warm season, the equatorial sun at midday can be truly oppressive. But throughout the day on the coast of Seymour a blue-footed booby colony remains vibrant with courtship. While blue-footed and red-footed boobies typically breed outside of the hot season, like penguins they are opportunistic and when food is plentiful the colonies are quick to take advantage. Couples court in a captivating ritual dance, sky-pointing with wing-tips and tails, passing token offerings of twigs. The blue feet, shining in the glare of the midday sun, appear almost unnatural, like the colouring in children's sweets. They show them off with carefully exaggerated footsteps as if standing in treacle. Females honk while males whistle, trying to impress. Walking among the unperturbed colony it is possible to appreciate what an attractive larder these birds would have seemed to the early visitors. The name 'booby' is thought to originate from *bobo*, the Spanish word for 'clown', perhaps on account of the birds' comical waddle, apparent stupidity and ease of capture. While some couples continue to court, others are already attending to their nest. 'Nest' is a generous description; it is nothing more than a scratched flat area, bordered by a radiating halo of jettisoned excrement, with a few twigs, bones and stones gathered during the tender ceremony of courtship. The parent sitting at the nest pants to cool down as it shades its clutch of between one and three eggs. Elsewhere on the rocky coast, the breeding of the Nazca booby, cousin to the blue-footed booby, affords us a harsh lesson in survival in the natural world. They produce two eggs, laid three to five days apart, but almost without exception only the elder will survive. The second chick serves as little more than an optional backup should the first egg fail to hatch or the elder chick perish early on. During their first days of life, the elder chick

monopolizes the food provided by the parent and will push its sibling out from the shelter of its parents' shadow. On the sun-scorched rock, away from the borders of the nest, the adult bird is no longer compelled to protect its younger chick and makes no attempt to rescue it, nor does it show concern when mockingbirds, finches and frigates descend to take it.

There is perhaps no better illustration of the bounty that these coastal waters can provide than the plunge diving of massed numbers of blue-footed boobies. Of the three species of booby in Galápagos it is the blue-footed that fishes the coastal waters and so is typically most observed plunge diving. At times hundreds flock off the coast and in unison fall to the sea, thrusting heads forward and pulling wings back in the final moments to cut into the water like darts. Such plunges propel the birds down to feed on shoals of small fish brought up to the surface by currents and the coastal topography. The lighter-weight males, distinguished from the females by slightly longer tail feathers and seemingly black pinpoint pupils, are more agile under water and find it easier to take off. They rise from the surface of the sea but quickly curl back to plunge again. It is an elegance and richness in stark contrast to their big-footed clumsiness on the rocky shore.

MANGROVE HAVENS

Standing out against the dark lava shores, glades of peaceful mangroves can be found lining stretches of the Galápagos coast, leading us into a different world. In their very design mangroves are adapted to find and settle on barren and raw coasts as found in Galápagos. They are true ocean travellers, with seedling pods tolerant of sea water and desiccation. Mangrove areas fringe sections of the coasts where sea conditions are more settled, wave action is less and bays and islets offer shelter for the seedlings to take root. Four species of mangrove are found along the shores of the archipelago, but none is endemic. Such supreme colonizers are less likely to form new species: the distance and severity of dispersal across the oceans and the hard coastal conditions of Galápagos are well within their capacities. So there is not the ecological pressure to adapt or change, and a relatively frequent arrival of new seedlings limits the extent of genetic isolation. Unlike many other species in Galápagos, on these exposed, barren coasts mangroves are in their comfort zone.

The mangrove glades are truly a world within a world, quiet and sheltered. In the shade of the red mangroves in Elizabeth Bay on Isabela, sea lions haul out on the branches, taking respite from the heat of the day. Moving back into the glades, the open sea feels increasingly distant. Turtles can be seen settled on the bottom, resting with only their heads extruding from beneath the blanket of warm sediment. Schools

Previous page: The enchanting courtship of blue-footed boobies. Males are most easily distinguished from the females by their seemingly pin-point pupil. The apparently larger pupil of the female is actually due to a dark ring in the iris, which the male lacks.
Opposite: A school of golden cow-nose rays drifts through the mangrove lagoon at Black Turtle Cove, on the north side of Santa Cruz.

of cow-nose rays drift through the lagoon and at times white-tip reef sharks can be seen gathering to mate and pup. It is one of those places in the coastal zone where the longer you look the more is revealed. For mangroves create a habitat and in doing so provide all manner of ecological niches to fill. An endemic lava heron is easily spotted, as it stands motionless, poised to strike at the juvenile fish that congregate among the prop roots. Great blue herons, common and cattle egrets also nest among them. In a few locations in the archipelago, if you are lucky, patient and know where to look, you may see the mangrove finch catching insects amongst the branches. Travelling through the channels, one can even see these colonizing plants being colonized. Beneath the water line, oysters and brittlestars have taken up residence on and among the prop roots. The regular ebb and flow of the tides through the network of the mangroves draw water rich with sediment and nutrients from the heart of the glades, providing a regular current of food for these filter feeders. Through these channels cruise Galápagos penguins. The mangroves serve as nurseries, harbouring the next generation of reef and open-ocean fish, and so are well stocked. The penguins know it and regularly blitz the glades in search of food.

But though rooted in place, the mangroves themselves should not be seen as static. In common with elsewhere in the world, changing sea levels and temperatures mean that mangrove cover varies over time in response to changing conditions. In Galápagos, however, there is another dynamic at play. Tectonic uplift can occur along the coasts in response to movements of magma beneath the surface. The most recent uplifts were recorded in 1954 and 1975 in the Bolivar Channel, where uplifting, by as much as 4 m and 70 cm respectively, caused mangrove glades to dry and die. The foothold of life on these islands is in constant review.

BEACH LIFE

The idyllic-looking beaches of Galápagos are more along the lines of what you might expect from an archipelago that straddles the equator. In places, white sands extend off shore, creating vivid turquoise shallows in Gardner Bay off Española or Tortuga Bay

off Santa Cruz. But you'll find no palm-lined beaches here, nor extensive networks of high dunes. The beaches nestle amid the lava and typically lead into scrub and low, dry vegetation. Coastal vegetation is characterized by plants that are designed to cope with salty conditions and little precipitation. The creeping vines of beach morning glory and mats of sesuvium help stabilize what dune systems there are. Though it occupies a relatively narrow band in the vegetation zones, the plant community here is highly prized by the coastal residents. Frigatebirds and red-footed boobies nest in the salt-bush, while sea lions and iguanas find shade and shelter where shrubs line the shore.

Above: Mangrove glades on Santa Cruz provide a sanctuary for many species on an otherwise barren rocky coast.

THE EASY LIFE

These beautiful beaches attract arguably the most charismatic resident of the archipelago, the Galápagos sea lion. Young sea lions appear perpetually at play. Their incorrigible sense of curiosity and mischief is captivating to watch. Pups from a month old gather in nursery pools that offer protection from sharks and orca, while adults can often be seen lazing on the beach. Though sea lions colonize many stretches of the coast and have more recently stamped their mark on the towns and harbours, beaches are

SPECTRUMS OF SAND

Seem from above, the beaches along the coast appear as oases amid a sea of lava rock. The sands are the product of the erosion of the rock fabric of the islands, combined with the broken remains of countless generations of shell and coralline-producing sea creatures. They offer a spectrum of colours depending on their position relative to the prevailing conditions. Whiter sand beaches tend to be found in the southeast of islands, where prevailing winds and waves accumulate greater amounts of marine material. Black beaches may result from the fall-out of tephra following eruptions, while those on more sheltered stretches of coast grow from the erosion of the land and take on the brown, red and even olivine-green colours of the surrounding rocks.

a particularly good place to observe them and the dynamics of a colony. Between June and December you may be lucky enough to witness a new recruit, typically born during the night or at first light. In the early morning frigates, mockingbirds and ruddy turnstones pick at the afterbirth. The baby is a dazed bag of wrinkles, but with a lot of room for growth, already instinctively clasped to its mother's teat, enjoying its first feed. The bleats of the newborn and the mother's calls and responses, along with their individual scents, form a tight bond of recognition. The cow will stay on land for the first four to five days in order to cement this bond and feed up the pup before returning to the water to fish. She may also continue to feed the previous year's offspring, so she must start hunting for herself as soon as possible in order to continue producing her rich milk.

These beaches also bear witness to some of Galápagos's most contested prize fights. For the largest male sea lions, the sand beaches, and the females they attract, are worth fighting for, and their reproductive future is at stake. During the breeding season between June and December, territorial clashes are common and, with the sea lion being the largest native animal of the islands, there is a lot of weight thrown around. Smaller, out-competed males are dramatically chased off, bolting down the beach and torpedoing through the shallows. Fights between equally matched males more commonly end with exhaustion in favour of one or the other. Although bites to the neck can draw blood, the thickness of the fat layers means serious injury is usually avoided. For these beachmasters, defending prime territories on beaches such as those on Mosquera Islet or Gardner Bay on Española is a full-time occupation. They can seldom afford to feed while holding dominance and therefore typically hold territory for only two to four weeks during the mating season. Retiring and recuperating beachmasters and younger males yet to rule gather in bachelor colonies on the less hospitable stretches of coast.

A falling tide draws out much smaller residents of the sands. Ghost crabs emerge from their burrows to sift through the debris left by the ebbing waves. Their

Opposite: Juvenile Galápagos sea lions gather in the shallows at Punta Suarez on Española; sea lions are encountered throughout the archipelago both on land and in the sea.

Above: A ghost crab feeds on the sand during low tide. Its multi-purpose pincers are used for fighting, to cut and tear larger items of food, and to manipulate sand grains into its mouth to feed on the micro-organisms these contain.

very existence here relies on the beaches, which offer both a home and a dining table. They methodically sort through the surface grains for micro-organisms, creating telltale sand balls in the process. But they'll also tackle larger items as they scour the wrack line. A Portuguese man-of-war jellyfish washed up on the beach makes for an opportunistic meal.

While adult sea lions travel out to feed, the young sea lions stick close to the shore. Perpetually playful in games of chase, like larger mammals throughout the world they are also honing skills that will help them survive when they reach independence. Marine iguanas returning from feeding offshore offer an irresistible target. They seem put out, but stoic, when their tails are tugged and chewed as they clamber back on land. Those sea lions that have already returned from fishing forays get back to lazing on the shore. But in their apparent laziness lies an important physiological condition. Like the marine iguanas, they have an oxygen debt to recover after having spent an average of 15 hours at sea feeding, diving in bouts to an average depth of 45 m. Given their aquatic prowess, the oceanic isolation of Galápagos – the greatest barrier for natural mammalian colonization of the land –

was simply not an issue for the sea lions that originally made their way down from the coasts of California to form this subspecies. But like other coastal creatures, sea lions suffer from the problems of El Niño years. It is those with the least reserves – the youngsters and large males defending territories – that suffer the greatest losses, largely from starvation. The rise of human fishing in these waters also brings increased risks and pressures for the sea lions. They run a gauntlet of nets, hooks and other marine debris. And there are reports of fishermen killing sea lions, those caught in nets being used as bait, and disturbing reports of sea-lion genitals being sold to the Asian aphrodisiac market. But seeing them slumber in the afternoon sun you'd be forgiven for thinking their life was easy!

The beaches of the archipelago offer sunbathing opportunities for another surprising marine creature. Green turtles drag themselves on to the sand to bask, or perhaps in the case of females to escape the intentions of breeding males. This behaviour is unknown in other places in the world, but is a trait of Galápagos green turtles. It was observed and commented on as early as 1697 by the naturalist–privateer William Dampier, who astutely observed that 'the turtle of these islands Gallapagos, are a sort of a bastard green turtle ... [and] different from any others, for both He's and She's come ashore in the day time, and lie in the sun.' They are of a distinctly dark form, *Chelonia mydas agassizi*, considered by some to be a separate species and found down the eastern Pacific coast from Mexico to Galápagos. Perhaps their dark coloration helps to optimize absorption of the sun's heat, the better to cope with the cooler, more productive waters.

This population of green turtles is by no means restricted to these beaches. Individuals tagged in Galápagos have been found in the coastal waters of Ecuador and Peru. But the females are known to return to the beaches from where they themselves hatched in order to lay their own eggs. Select beaches across the archipelago offer the necessary quality of sand required to excavate the nest and incubate their precious clutch of eggs. On spring tides during the nesting season, the females lumber out of the water. They haul themselves above the high-water mark to dig a pit, in which to lay their clutch of 70–80 eggs. In the peak season, the nesting of numerous females creates a crater-ridden beach, with their telltale tracks back to the sea left in the morning sand. Fifty-five days later, usually under cover of darkness, the hatchlings emerge and run the gauntlet of herons, frigates and ghost crabs to reach the sea.

NORTH AMERICAN VISITORS

Though we often think of Galápagos as an emblem of isolation, for some species these islands are just a stopover on a regular world tour. Despite their remoteness there is migration to Galápagos and its beaches and lagoons are a particularly strong draw. On most beaches around the archipelago sanderlings can be seen scurrying back and forth with the ebb and flow of each wave, searching the beach for intertidal molluscs dug into the sand. The long, curved bill and chopstick legs of a whimbrel are easy to identify. It has travelled from breeding grounds in the Arctic to spend the northern winter on the beaches and in the lagoons of Galápagos. Above the crash of waves the telltale call of a wandering tattler can be heard, a song also associated with the higher mountain streams of northwestern America. As most tourists will agree, the beaches of Galápagos and the unique niche that the sands provide are worth travelling for.

In a handful of places across the islands, the beaches and mangroves lead into salt-water lagoons. Such lagoons not only provide a habitat for many of the coastal waders and birds associated with the beaches but also harbour a few surprises. In locations on Santa Cruz, Isabela and Floreana, greater flamingos can be seen filter-feeding sediment to collect the water boatmen, shrimp and other small crustaceans that flourish in the shallow water of the lagoons. But it's far from a quick snack. They must typically feed for up to 12 hours a day to gain sufficient food, sifting the sediment with plates in their bills that catch the organisms as they are pumped through. The carotenoid pigments in the crustaceans are partly responsible for the birds' characteristic pink colour, the pigments being broken down in the liver and transported to the feathers, skin and even egg yolk. The Galápagos flamingo population is small and regularly moves among lagoons, seeking out the best conditions. But unlike so many Galápagos residents, the flamingos, whose ancestors arrived comparatively recently from the Caribbean, still harbour a fear of man and are easily disturbed, so must be watched from a distance. Their natural wariness is a reminder of how unaffected the majority of Galápagos wildlife is by human presence, and how accessible, and a privilege, wildlife watching is here.

LIFE ON THE CLIFFS

While the sand beaches illustrate the continual erosive power applied by the sea on the archipelago, it is at the cliffs that the full force of the water can be seen. Standing on the cliff tops of Genovesa or Española, watching an oceanic swell rolling landwards, rising into waves that curl up the cliffs or break against rock, one can appreciate the huge forces at work. At Española's blowhole, these forces are framed in a visual spectacle of dynamic beauty. The cliff tops, the flat plateaus and, in places, cracks and crevices formed in the lava create another unique habitat of which many types of seabirds have taken advantage.

The medley of different species of seabirds found in Galápagos becomes apparent in these cliff habitats. They are all tapping into the same general resource – the productive waters surrounding the islands – but to do so they must separate their activities in space and time. The three species of booby, for instance, avoid competition by separating their fishing grounds and the different fish each provides: the blue-footed booby fishes the inshore waters, the Nazca booby opts for the middle ground, usually between islands, while the red-footed booby ventures furthest offshore. For the red-footed booby, the effort to fish further offshore and the time it takes to return to the nest with food impacts on their breeding strategy, forcing them to take longer to raise their single young and drawing out the breeding cycle to over a year.

The range of fishing techniques and strategies employed amongst the seabird communities to coexist becomes evident during a trip through the islands. Most distinctive must be the crowd-pleasing brown pelicans, often encountered plunge diving to engorge their characteristic pouch with gallons of water, which is then drained, leaving its catch. During feeding frenzies offshore, they can commonly be seen accompanied by brown noddies, Galápagos shearwaters and storm petrels, which pick morsels from the sea or also dive for fish.

Of all the seabirds to colonize the cliffs of Galápagos, the most majestic is without doubt the waved albatross. Only the cliffs of Española offer the take-off and landing strip required by these largest of Galápagos birds. So specific is this requirement that they are not only endemic to the archipelago but also considered endemic to this island, though a number of pairs also breed on Isla de la Plata, off the Ecuadorian coast. For up to six months of the year they stay out at sea, scavenging and fishing the ocean surface for squid and plankton. They journey over the vast Pacific, but return to their home cliffs of Española to breed. From mid-April to mid-December the colony comes alive as the males return to their nest sites first, followed a few days or weeks later by their lifelong partners. They are stunning-looking birds with a beautiful charcoal-grey graded breast plumage and curiously poignant eyes, and their courtship is one of the most beautiful scenes to be witnessed in Galápagos: a show of jousting, bill-tapping and sky-pointing.

Elsewhere in the islands similar cliff habitats provide shelter and protection for the albatross's smaller relatives, the storm petrels. These are some of Galápagos's daintiest avian residents and, like their larger relatives, they too feed from the surface of the sea, skipping among the waves, picking at plankton. Their small size means that in places such as the cliffs of Genovesa, Castro's and Galápagos's storm petrels are able to nest within the network of small lava fissures and cracks. But their size also puts them at the bottom of the pecking order of the seabird community, leaving

Above: A male great frigatebird displays at its nesting area on Genovesa. This spectacular courtship can go on for a number of weeks before mating begins and a single egg is produced. Overleaf: A brown pelican in the early morning on the beach at Quinta Playa, southern Isabela.

them particularly vulnerable. The lava network offers protection from aerial assaults, but it can also conceal an ambush predator. Uniquely, on Genovesa, in the absence of the Galápagos hawk, the resident short-eared owls have taken to hunting during the day. They lie in wait, then grab at a passing storm petrel, returning to its nest.

Frigatebirds pose another deadly threat to the storm petrels. Infamously piratical, frigatebirds are commonly seen following tourist or fishing boats or chasing boobies returning from fishing offshore, forcing them to regurgitate their catch, which they then snatch. But to the petrel colony the frigatebirds can be lethal. Lurking on the thermals and updrafts, silhouetted against the sun, their scissor-shaped tails, sculpted wings and ominous beaks create a fearsome outline. They will intercept the storm petrels returning to the colony with deadly effect, catching them and even eating them on the wing. By way of avoiding such threats, the larger endemic Galápagos petrel has opted to travel in the hours of darkness between their burrow nests in the highlands and their offshore fishing grounds. In a haunting chorus that fills the crater of Cerro Pajas on Floreana and Media Luna on Santa Cruz, they can be heard returning just after dark, then leaving again before daybreak.

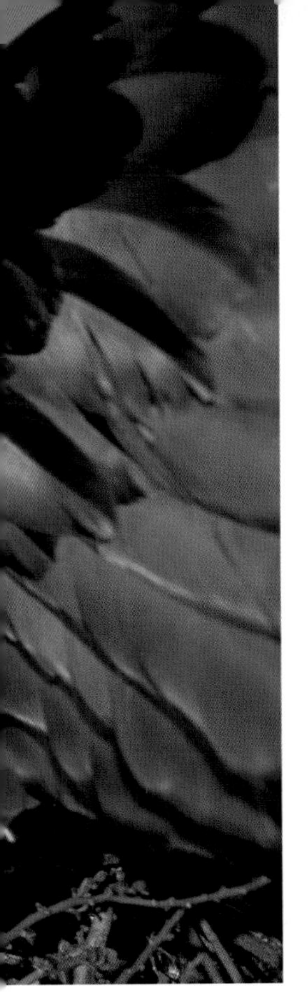

In the busy coastal seabird communities, vivid and noisy breeding displays are typical of many species adapted to colonial living. The dances of blue-footed boobies and waved albatross have already been mentioned. But there are others. The screaming display calls of the red-billed tropicbirds are as unmistakable as their distinctive elongated tail plumes and vivid red beaks. These are true cliff-dwellers, colonizing ledges and cracks from where they launch their aerial display – they head out to sea, then circle back to the cliff face, flying in bursts of fluttering and gliding, letting out their shrill cry. But the spectacle not to be missed is the display of male frigatebirds. Galápagos is home to two species of frigates, the magnificent frigatebird and the great frigatebird. The female great frigatebird can be distinguished from the female magnificent frigatebird by the distinctive red eye ring and white breast feathers that extend past the throat and up to the chin. It is something that the males of each species must also watch for as they choose which passing female to display to. Males position themselves on a nesting site among the salt-bush scrub or palo santo, balloon out their bright red gular sac and let out a melodic gobbling call to attract a potential mate.

Among the diverse selection of seabird found on Galápagos's cliffs, one of the most intriguing residents is the beautiful endemic swallow-tailed gull. It provides one of the best examples of niche separation of a seabird species by notably opting for the night shift. These gulls spend the hours of darkness cashing in on the rich bounty of squid and plankton that vertically migrate to the shallows during the night. Their large, black eyes, rich in low-light vision 'rod' elements and bordered by red rings, have adapted to allow them to fill this nocturnal niche. They can locate the phosphorescent glow of food in a way that diurnal species such as the red-billed tropicbird, which fishes in the same waters, can not. Back at the nesting sites on the rocky cliffs, in the darkness, a chick taps at the white marking visible on the tip of its parent's beak to release the rich concoction of squid and fish.

But where are the feeding grounds to which these gulls are heading? Do they travel to upwelled waters above the seamounts of old, submerged islands? Or perhaps they source the productive convergent zones between currents? Maybe they fish in the wake of whales, feeding on the trail of life and debris left behind. Despite its being such an insightful laboratory for ecological and evolutionary science, there still remains much to be explored about the workings of life in Galápagos. And many of the mysteries lie beneath its waves, the subject of Chapter 6.

6

The very *lifeblood* of Galápagos's marine world lies in the diverse ocean currents that converge on their remote shores. *Countless species* have ridden these *ocean highways* from as far as the subarctic, the tropical reefs of Panama and the Polynesian islands of the western Pacific. This dynamic ocean crossroad has ensured a *unique oasis* of marine life: only in Galápagos do cold-water penguins chase through tropical-reef fish to catch their cold-ocean sardines.

OCEAN OASIS

Previous page: In non-El Niño years, Galápagos sea lions hunt less and have more leisure time. This one may be chasing a school of endemic black-striped salema simply for fun. Above: Common dolphins are among the most frequently sighted cetaceans in Galápagos waters. Leaping from the ocean adds the turn of speed necessary to keep up with the schools of tuna fish on which they prey.

A BOUNTIFUL SEA

Unlike the desolate appearance of life on land, the prolific marine life of Galápagos seems to have impressed visitors to the islands for centuries. Gaze over the ocean and it will soon hint of profusion. Beneath the surface, shadows are always on the move. Perhaps an escort of male turtles swims in hot pursuit of a female as she surfaces to gasp air. A sinewy shape meanders by until a dorsal fin cuts a subtle wake confirming it is a shark. The frequency of sightings is also impressive. A school of 20 or more cow-nose rays may glide gracefully past, or perhaps a pod of sperm whales will doze peacefully on the deep, smooth roll of open ocean. The calm that prevails at one moment is often shattered as a flurry of diverse marine predators converges on an ephemeral aggregation of bait fish. Dolphins, sea lions, killer whales, tuna, manta rays, billfish, sharks, seabirds and others congregate in the desperate chaos of brief opportunity. Such spectacles of marine mega-fauna are still quite common in Galápagos and are supported by a cast of marine life that ultimately depends on the great abundance of microscopic phytoplankton (algae) that are the foundation of the food web. These are rich and diverse waters but, above all, they retain much of the glory lost from other formerly comparable areas around the globe.

A SEA OF CONFUSION

It soon becomes evident, when diving in Galápagos, how variable and dynamic oceanic conditions are here. In the course of a single dive, a warm clear-blue ocean can be replaced without warning by a frigid, green soup, accompanied by a host of different creatures. Temperatures may change by 5°C in just a few minutes, and the ocean, calm and tranquil one moment, may surge with strong currents the next. Generally, Galápagos's marine realm is categorized as 'subtropical', but this classification, as with so many other things in these islands, is debated; at the moment there is not firm agreement as to how many temperature zones actually exist in Galápagos waters. The forces that drive this dynamic complexity are influenced by the convergence of two distinct oceanographic regimes, one in the north, the other in the south. The waters surrounding the northern archipelago are warm and thus can be classified as 'tropical'. In the south, however, the ocean tends to be far colder than one would expect so close to the Equator, so the region can be classified as 'warm temperate'. Consistently cool temperatures also prevail in the large western islands of Isabela and Fernandina. The influx of deep-water nutrients make this region the most productive in the archipelago.

Hosts of Sally-Lightfoots were the most brilliant spots of colour above the water in these islands, putting to shame the dull, drab hues of the terrestrial organisms and hinting at the glories of colourful animal life beneath the surface of the sea.

William Beebe,
Galapagos: World's End (1924)

UNDERWATER 'WINDS': CURRENTS IN PERSPECTIVE

The Galápagos islands lie between two climatic extremes. Some 550 km northeast, Cocos Island (off Costa Rica) is one of the wettest places on Earth, a classic palm-clad tropical island with waterfalls cascading directly from the verdure into clean, warm seas. Just south of Galápagos, the Paracas Islands (off Peru) are subject to persistent drought and their landscape is dominated by desert. The contrast can be explained by the fact that Cocos is surrounded by warm oceans, which ensure plentiful rainfall, while the Paracas are engulfed by the cold Humboldt current and its associated rainless skies. Galápagos falls between these extremes, or rather mixes them. Both cold and warm ocean currents converge here, but they dominate at different times of the year.

In normal years, the archipelago receives a westward surface flow of ocean called the South Equatorial current. It originates off the coast of Ecuador and flows past Galápagos on its way to Polynesia, in the central Pacific. Between May and December, southeasterly trade winds strengthen the cold, nutrient-rich waters of the South Equatorial current. This current system brings with it species from Chile and Peru. Penguins, fur seals and other cold-water species are thought to have hitched rides to Galápagos via this current. These species survive here because there is a second source of cool, rich water of western origins.

The westward flow of surface water drags a deep undercurrent in the opposite direction. The Equatorial countercurrent or Cromwell current travels eastward, across the Equator, deep in the ocean until it collides with the Galápagos Platform on the west side of the archipelago. The steep 4000-m volcanic slopes cause the current to surface rapidly with cold, nutrient-rich water welling up near Fernandina and Isabela. This cool sea then wraps around the archipelago, causing local upwelling off several other islands. Species more commonly associated with temperate and and even sub-polar affinities owe their continued existence in Galápagos to this highly productive undercurrent.

Above: Great travellers in the larval stage, Moorish idols grace the waters of many isolated Galápagos islands. Opposite: Like the terrestrial wildlife, many Galápagos fish have a reduced fleeing instinct. Despite increasing and illegal spear-fishing, these yellow-tailed grunts still allow divers to get close enough to study their brightly coloured markings.

From around January to April the South Equatorial is warmed by the southerly migration of the sun towards the Equator. The trade winds shift and lighten, becoming north-easterly and pushing the warmer (26–28°C), nutrient-reduced Panama current into Galápagos. Most tropical species arrived here via the Panama current.

As if that were not complex enough, catastrophic events such as El Niño (see Chapter 1) also have an important role to play. The Moorish idol is a graceful yellow and white fish, with long dorsal streamers and a finely elongated mouth. Its presence here is surprising, considering that its evolutionary homelands are in the tropical reefs of the western Pacific, thousands of kilometres away. For it to have

A PROMISED LAND: THE BILLION-DOLLAR BIRDS

On the dry Paracas Islands, the cold, nutrient-rich sea causes millions of seabirds to deposit 'guano', which builds up in dense accretions. The cold ocean also ensures the guano is seldom washed away by rain. This excrement is rich in nitrates, a key ingredient for the making of explosives before modern technology took over. During the 1800s nitrates became so valuable that a 'white gold rush' ensured the rapid depletion of guano on Paracas. In the 1830s it was assumed Galápagos must also have guano beds, and the Ecuadorian government nearly persuaded the USA into an expensive agreement for exclusive rights to export guano. Unlike Paracas, however, Galápagos does not sustain millions of guano-producing seabirds and the insignificant deposits that do exist are washed away by the warm ocean-induced rains.

Above: Most of the Galápagos penguin population is concentrated on the shores of the far western islands of Fernandina and Isabela. But a few of these tuxedo-clad birds can be found mingling with snorkellers off the shores of the central islands.

arrived in Galápagos, a surface current all the way from those hot waters must have made it to the eastern Pacific, countering the normal direction of their flow. Given observations during recent El Niños of the sporadic occurrence of Indo-Pacific species, a plausible explanation for the establisment of the Moorish idol is an El Niño of devastating proportions. The nutrient-poor water would have strangled the islands' productivity for as long as it lasted, causing many individuals to starve and some populations to go extinct. Like the poppy that blooms on a silenced battlefield, the Moorish idol reminds us of past calamity.

ARRIVING BY SEA

Seeing penguins, a cold-water icon, in a maze of tropical red mangroves or darting among coral heads is counterintuitive: you don't expect one place to cater to species from polar-opposite climates. Yet in Galápagos it is quite normal to see creatures whose ancestors originated in very different places co-existing under one ecological roof.

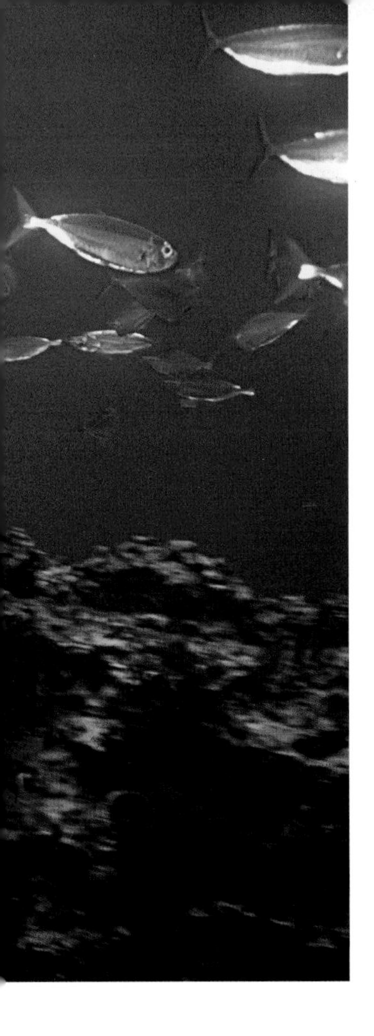

One would expect that originating closer to Galápagos would directly increase a species' likelihood of colonizing the islands, but scientific data suggests it is more complex than that. A species is more likely to become established in Galápagos when its breeding season in a particular area coincides with a favourable flow of currents towards the islands, and it happens that the Panama current flows to Galápagos just at the peak reproductive season for the Panamanian marine fauna.

Another factor is that the longer larvae can remain cryptic as zooplankton, the more possibility they have of avoiding open-ocean predators. Organisms with a prolonged larval stage have a better chance of arriving in Galápagos. For example, moray eels are masters of ocean travel and account for 22 species here. But marine organisms from the southwest coast of South America live in a very narrow band of shallow ocean that gives way to a deep abyssal drop-off. Their larvae change very quickly so that they can swim back on to the shallow continental shelves before they drift into the abyssal canyons and are lost for ever. These poor travellers from the Peru/Chile province account for only 7 per cent of the Galápagos marine system.

Sometimes, no matter how mobile a species is, the barriers to dispersal are insurmountable. The ancestor of the Galápagos black-spot porgy originated in the Atlantic and was totally cut off from its parent stock when the Panama isthmus rose up to separate the Atlantic from the Pacific. Any previous genetic flow between the two populations ceased completely, leaving the Galápagos porgy to pursue an entirely independent evolutionary history.

A DIVERSITY OF HABITATS

Species that establish in Galápagos tend to settle in habitats that most resemble their ancestral home. Cold-water species generally prefer western Isabela and Fernandina, or the southern islands of Española and Floreana. Many arrive to find a wide range of real estate available but not all newcomers are welcome. For species originating in shallow coastal tidal flats, Galápagos offers meagre pickings. The steep, rocky terrain and limited tidal range do not encourage the formation of this habitat.

With that exception, weathering and organic interaction with lava have

Above:
The hieroglyphic hawkfish's beautiful patterns act as ocean camouflage as it ambushes fish and invertebrates. Its disguise is so effective that one scientist even reports seeing a hawkfish capture a marine iguana!

created an unusually wide variety of habitats in Galápagos. Most of the coastal shores are sloping lava reefs interspersed with sand derived from coral and echinoderm exoskeletons mixed with pulverized lava. Coral reefs, vertical walls and sand flats intermingle with the lava reefs to produce a biodiversity maze.

ROCKY BOTTOM

Under water, the flanks of the volcanoes are strewn with collapsed boulders and rocks – just perfect for invertebrates that prefer to remain hidden from view. At first glance, despite some scattered fish 'hot spots', the ocean floor appears rather empty. It is hard to get an accurate sense of this habitat when the vast majority of creatures are hiding under rocks during the day or moving too slowly to be noticed. In contrast, night-vision and time-lapse cameras reveal these rocky slopes as frantic rush-hour highways. Sea cucumbers, brittlestars, urchins, crabs, lobsters, shrimp, octopus, fire worms and many others cruise about the reef in the scramble to make a living. Sessile creatures that are attached to the reef, such as cup corals and anemones, open their polyps and actively feed at night. There are larger creatures about, too. Moray eels slither in and out of holes, sensing their way to the next meal. Port Jackson sharks are often found close to the bottom, well camouflaged by leopard spots, their snubby faces ideal for eating invertebrates from the sea floor.

As daylight begins filtering though the ocean, the creatures of the night slip away before a new cast of characters appears. Close observation reveals flamboyant invertebrates with some very interesting behaviours. The colourful, fleshy opening of a magnificent scallop puts the neon signs of Las Vegas to shame. Sensors located in the scallop's mantle warn them of impending attacks. When predators such as some sea-star species get uncomfortably close, scallops can make surprisingly rapid, shell-snapping, jet-propelled emergency exits.

Small nudibranchs are also neon signs in the gloom. But though humans use sight as their primary sense down here, many marine creatures have more refined electrical, pressure and chemical senses. The endemic blue-striped nudibranch leaves a telltale trail that a carnivorous relative, *Roboastra*, may pick up, following the scent and devouring its prey in one mouthful. People have also taken a dietary liking to marine invertebrates and one such animal has changed the fate of Galápagos for ever (see box, page 149).

The rocky bottom is also home to colourful swathes of tropical reef fish that flit among rocky outcrops. Yellow-tail surgeon fish are ubiquitous, ganging up in algae-feeding frenzies as they work the reef. Some people think the scalpel-like protrusions from their tails (from which their medical-sounding name is derived) may serve as a slashing 'weapon' to ward off other fish. But neither their numbers nor their defences seem to worry one reef dweller.

The yellow-tail damselfish lives alone, tending its patch of algae with Herculean determination. No army of fish or pelagic giant will deter this pugnacious little creature from defending its territory. Even white-tip reef sharks attempting to rest on the bottom are attacked. Its aggression is also matched by surprising practicality. The coral- and algae-eating pencil sea urchin competes with the damselfish for living space and food, but is heavily armoured against attack. The resourceful damselfish deals with encroaching sea urchins by simply picking them up by one of their spines and swimming them off the premises.

This ingenious strategy may have profound ecological implications. An interesting feature of islands that have never been connected to the mainland is the tendency for species to change their behaviours due to shifting ecological conditions. For instance, on mainland Ecuador the pencil sea urchin has many more predators and grows cryptic and small to avoid predation. In Galápagos the presence of fewer predators has enabled these urchins to grow larger and to diversify their diet. Since corals are to their liking, studies near Champion Islet, near Floreana, suggest they

CUPID'S PRICE

While they don't look anything like sea urchins or sea stars, sea cucumbers are members of the *Holothurian* family, in the same taxonomic class – echinoderms. As a group they are not as well known as their relatives, probably because most species hide under rocks during the day. At night they slither over the reef to feed and, as they go, recycle nutrients and oxygen back into the substrate. Their ecological role, similar to that of earthworms in the soil, enhances the productivity of the sea-floor community.

Asian and Indo-Pacific countries prize sea cucumbers as a culinary delicacy, more for their reputed aphrodisiac powers than for either taste or texture. This reputation fetches large sums. High demand has caused a fever of sea-cucumber fisheries to open around the world. In 1988, Ecuador joined the ranks as a major exporter and a few years later illegal exploitation began in the Galápagos Marine Reserve. This nondescript echinoderm caused a short-lived 'gold rush' and the effects had parallels with the frontier days of the Wild West. The size of the fishing fleet doubled overnight, countless strikes and riots were instigated by fishermen to force authorities to extend fishing seasons, a park guard was shot, and several people and a giant tortoise were taken hostage.

The tumult of the sea-cucumber fishery changed the socio-economic and political structure of the local community, pitting many people against the National Park Service and undermining its authority. Today the fishery is in tatters with the sea cucumbers virtually commercially extinct, an ecological loss and a sad legacy of years of political manipulation and weak government.

may have locally destroyed up to 30 per cent of corals. Researchers have found that where damselfish inhabit the reef urchins have less impact on coral.

SAND FLATS

Eroded coral, marine-organism exoskeletons and organic sediment drapes a smooth, soft layer of sand over the brittle rock and creates another important habitat, especially for burrowing creatures. The sand can be interspersed among rocks or found in large tracks of uninterrupted flats.

Many creatures that do not bury themselves in the sand adopt cryptic forms that blend into the background. As sand varies in colour around the islands, so too do members of a species living in different-coloured habitats. An example is the odd-looking red-lipped batfish, whose populations are much darker over brown volcanic sand than over white coralline grains. Sea lions love a good game of underwater volleyball and take comic advantage of these slow-swimming batfish to serve as their 'ball' and occasionally as food. This adaptive change may limit their predation.

But adapting to your surroundings is not just about defence. The batfish

Opposite: Sea lions play in collapsed lava tubes; these lava grottoes provide a refuge from predators such as sharks or orca whales.

Above: Christmas-tree worms unfold their colourful spiral fans as they feed on plankton. Their barbed tube shells are excellent armour against predators and empty tubes make secure homes for other creatures, such as tube blennies.

may also benefit by blending into the background to hunt prey more effectively. Some creatures could not be more conspicuous over the sand and need to worry less about predation. Sea stars rely on their sturdy protective exoskeletons that allow them free rein over the open terrain. Others, like the Pacific razorfish, live by their wits, raising a spiked dorsal fin to ward off predators or, when all else fails, darting under the sand to hide.

The sand acts as a useful 'smokescreen' for other species also. Anemones are relatives of the corals; to defend themselves they do not secrete calcium around their bodies as corals do, but rather retract into the sandy substrate of the flats. Like the corals they also use their toxic tentacle tips to catch plankton from the water column.

In fact, though the sand flats look like an underwater desert, the habitat is brimming with life. Among the grains you will find a host of species from polychaetes through gastropods to crustaceans and other species. They are an important source of food for animals that rely on the sand habitat for much of their diet. Many ray species feed here and exploit different niches. Long-tailed stingrays often leave a tell-tale sign that they have been feeding – they excavate bomb-like craters in order to find crabs and clams. The eagle ray, one of the most serene animals on the reef, also feeds on the sandy bottom. Using its duckbilled mouth it detects the minute smells and electrical currents of molluscs, crustaceans, shrimps, crabs, octopus and worms buried in the sand. Eagle rays are not confined to the sand flats and swim gracefully throughout the shallows; they also visit the rarest of all Galápagos habitats, the coral reef.

CORAL REEFS

In most areas around Galápagos, you will notice a dearth of any significant coral reefs. The cold currents sweeping into the islands half the year inhibit coral growth and only small outcroppings of the 22 or so species occur, generally in waters protected from consistent cold upwelling. Corals prefer warm water, so the largest reef-building corals are found in Wolf and Darwin islands in the far north of the archipelago.

EL NIÑO: CAN WE MAKE A DIFFERENCE?

A very real threat of extinction hangs over many Galápagos marine species. Indeed, several residents have already vanished. The blackspot chromis went extinct as a result of two consecutive warming events in the 1970s. After the 1982–3 El Niño, others like the rusty damselfish vanished from Fernandina, where its algae habitat has barely recovered – the fish hangs on with vestigial populations in a few areas of Isabela.

El Niño stress on the environment is compounded by fishing. An aggressive lobster fishery has significantly reduced populations of spiny and slipper lobster and it is postulated that with fewer lobsters around to prey on sea urchins, the latter's populations have exploded. The urchins are now eating the remaining strands of El Niño-reduced coral, pushing these vulnerable species closer to extinction. When El Niños already stress the environment to the limit, human demands for resources are an additional pressure that could up the scales to extinction. We are a fundamental cog in the ecological mechanism, so looking for ways to offset our impact during El Niño years may reduce the risk of other species following in the fin print of the blackspot chromis.

HAMMERHEAD SHARKS: PREDATORS AS PREY

Hammerhead sharks are thought to have appeared in the Miocene epoch roughly 23 million years ago. Their ancient cycle of life pre-dates the current islands: seeing these sharks now is like witnessing a spectacle from an antediluvian sea. In certain places they emerge out of the blue in numbers that astound the senses. Wall upon wall of their bodies glide by in majestic silence, reflecting a mesmerizing hue of metallic grey until they gradually fade into the distance.

Hammerheads spend much of their time cruising open ocean, but they also have a close association with the reef. As they glide into the shallows, some peel off from the main school and head for concentrated pockets of fish. But their approach is not predatory – they are just looking to be cleaned. In time-tested ritual the sharks invite these reef dwellers for a meal and in return they swim away less burdened by parasites. Occasionally the fish are so enthusiastic to feed that parts of the hammerheads are lost behind a veil of glittering scales. The symbiosis is never more clearly illustrated than when the delicate reef fish pick off parasites around the open mouths of these powerful predators.

With spectacles such as this, it is little wonder that tourists fly thousands of kilometres to dive with the Galápagos sharks. Every year, shark tourism generates millions of dollars for the Ecuadorian economy and supports tourist-related livelihoods. This phenomenon is indeed another symbiosis, as tourism and ecology work to keep sharks in their place as top predators. But the relationship is threatened by a competing industry. The Asian demand for shark-fin soup has catalysed a worldwide commerce in shark finning. It's an ugly business – the shark is pulled aboard the boat just long enough for its fins to be sliced off, then is often thrown back into the sea to die. The short-term profits of finning often eclipse the long-term benefits of shark tourism. In some areas of the world, shark populations have decreased by more than 90 per cent and, sadly, the same fate is possible for the sharks of Galápagos. Several environmental organizations have emerged to combat the effects of illegal fishing and ensure the future of these ancient creatures. If fishing and the demand for fins is not curtailed, the prospect looks bleak. Animals that have been swimming since before the dinosaurs will have vanished from the face of the Earth and our children may never experience the thrill of merging imagination with an extraordinary reality. While we may survive without such esoteric luxuries, the real question is what will happen to the whole ecological system on which millions of people rely for protein, if we remove a keystone species from the apex of the food chain?

WHALE PARADISE

During the whaling days of the 1800s the Galápagos fleets focused their slaughter on the western islands. Today satellite images from space show that the thickest algal blooms are in a very narrow band corresponding to the length of Isabela, where cetaceans tend to predominate. The limited feeding range in this area made whales an easy target and it is not hard to see how this contributed to rapid declines in their populations. Over 200 years on, the protected waters around Fernandina and Isabela still harbour the survivors of this holocaust. Cameras and binoculars have replaced harpoons and it is common to hear or see whales. Thanks to the latest research tools and some remarkable oceanography, we now know that the islands play a greater role in encouraging this spectacle than was previously understood.

It is thought that the concentrated nutrients the Cromwell current brings to the western islands are caused by the Galápagos Platform forcing the deep ocean undercurrent into the sunlit shallows. When satellites were trained on Galápagos, the areas north and south of the islands generally showed up as an impoverished ocean. This was to be expected in the north, where warm waters are associated with low productivity. But it was a surprise that the cold ocean south of Galápagos also exhibited little productivity, despite a reasonable nitrate content.

The satellite showed that the only areas with consistently high levels of plankton were around Fernandina and Isabela. The late oceanographer John Martin wanted to discover why that area was special. He found trace samples of iron around Fernandina, which led him to suspect that this mineral was a key element in the production of algae. Armed with an educated guess and a boat-load of iron filings, he set off 800 km south of Galápagos to test his hypothesis. He managed to instigate a huge plankton bloom that proved iron was the missing micro-nutrient needed to generate the production of algae. We now think that iron leached from the lava of the Galápagos Platform by weathering and erosion is one of the key ingredients in the high-productivity algal bloom around the islands that the whales so enjoy.

Many fish use the reefs for protection from predators in the surrounding ocean, but the coral is also subject to predation by numerous species. Telltale white teeth marks indicate where fish like the blue chin or bumphead parrotfishes have gnawed away at algae and inadvertently destroyed coral polyps in the process. Invertebrates like the pencil sea urchin and crown of thorns starfish also prey on coral.

Numerous coral beds in Galápagos are thought to be 2000 years old and one patch near southern Isabela has been dated at 6000 years. It is only in the last 25 years, since two events of super ocean heating from powerful El Niños, that 99 per cent of all coral in Galápagos has disappeared. Diving in Wolf and Darwin you would not suspect that you were in what makes up 95 per cent of the remaining coral in Galapagos.

The greatest diversity of fish in the Pacific is somewhere near Fiji, thousands of kilometres away, so you won't see too much fish diversity on Galápagos coral reefs. But what this outpost of marine life lacks in diversity, it makes up for in sheer numbers. Barberfish can school in hundreds, speckling the reef with their glittering yellow bodies. While the reef serves to give them protection from predators, part of their livelihood is actually based on attracting some formidable predators to the reef, as the box on page 152 explains.

Right: The Pacific seahorse is the only seahorse in the eastern Pacific, often found on rock walls with its tail twisted around gorgonians or black coral. Male seahorses with extended bellies have not overeaten – their pouch is full of eggs, which they brood until the hatchlings disperse.

The water column, the space between the surface of the sea and the ocean bed, is often scattered with Creole fish. The ember-like glow of their crimson bodies is a ubiquitous presence in the ocean and a good sign that plankton-rich waters are around. These fish are plankton feeders and normally establish themselves where currents are accentuated. Their frantic feeding is a sign that an upwelling of nutrient-rich ocean has initiated a bloom of algal phytoplankton. Algae are the essential staple that kick-starts the food chain around the islands.

WALLS AND LEDGES

Further down, the radiant crimson glow of Creole fish gradually fades into a dusty brown as the red spectrum of light is absorbed into the water column. The reef appears as a series of serrated undercut ridges and ledges, descending out of sight into the depths. Each line of undercut rock indicates a period of lower sea levels when the waves eroded soft volcanic ash into serrated ledges. The ash substrate is pitted with holes and crevices that provide safe homes for a myriad of reef-dwelling creatures hiding from opportunistic predators.

Endemic black coral, when alive, actually has bright yellow-green branches. It roots in the soft substrate and often dominates the habitat. Its elegant strands in turn attract many other creatures for protection. The long-nose hawkfish calls attention to itself with its geometric red lines and green eyes. Not far off, its cousin the coral hawkfish has a quite different, snub face. It makes one wonder if these fish aren't separated by their feeding apparatus as surely as the finches are (see box, page 155).

Underwater walls are a common feature of volcanic islets. Tens of metres down, attracted by the high productivity of upwellings, an abundance of invertebrate

DARWIN'S FISHES

Left: This yellow variety of Aulostomus chinensis *grows to 80 cm in length. Crafty hunters, they often use other fish for cover to launch their streamlined attacks. They can change colour quickly – a useful strategy for hunting and escaping the notice of larger predators.*

Before Charles Darwin arrived in Galápagos in 1835, not a single fish had been collected or described from there or anywhere in the eastern Pacific. He collected 15 species of fish, which he preserved in spirits for the journey home. What he learned from these collections resulted in the first bio-geographical notes on the Galápagos marine environment.

Darwin's keen interest in Galápagos fish is detailed by Dr Daniel Pauly in his book *Darwin's Fishes*. Dr Pauly believes that Darwin gained an awareness of speciation from his observations of Galápagos fishes as surely as he did from the creatures on land. What Darwin realized was that islands situated close to other landmasses normally had a strong representation of fish from those landmasses, whereas islands far away from the mainland 'possess an almost entirely peculiar fauna' and that in Galápagos 'sea shells and even fish, are almost all peculiar and distinct species not found in any other quarter of the world'. Of the 15 fish Darwin collected, five were unique to Galápagos. One, the Galápagos sheephead, *Semicossyphus darwini*, was named after him.

life clings to these walls. From a distance the wall is a three-dimensional tapestry of earth tones, with the graceful, brown fans of gorgonian corals dominating attention. Radiating from thousands of tiny projections, their delicate tentacles search for plankton. Like most invertebrate life attached to the rocks, they derive their living by filter feeding.

Past the gentle undulations of gorgonians, acorn barnacles usually stud the wall in tightly packed formations. Barnacles are actually arthropods most closely related to crabs, prawns and lobsters, and may have proportionally the largest penis-to-body-length ratio in the animal kingdom. To breed they will extend their appendage and search out other barnacles in the vicinity. Barnacle larvae go through several distinct swimming stages as plankton, before settling head down on a rock. From there they spend the rest of their days upside down with their fan-like legs sweeping the water column for food. It's a good living: giant barnacles can multiply their biomass 16 times in one year.

OPEN OCEAN

Not far from the wall the current sweeps rapidly into deep, open ocean. Here the sun flickers into the depths, illuminating the creatures that inhabit this enormous expanse of sea. Sometimes it is scarcely perceptible, but the ocean buzzes with life. It takes a microscope to detect the activity of zooplankton and phytoplankton fighting tiny battles of predator and prey in the void. As you adjust your sight to the scene, hidden creatures gradually emerge. Most are translucent or transparent to take on the light behind them – as effective a camouflage as you can imagine. Jellyfish, tunicates, sea squirts and a host of other pelagic inhabitants waft at the mercy of the currents. Some of these ocean travellers are islands to themselves, providing shelter for hitchhikers such as shrimp and tiny fish. For those without a floating castle, being small and inconspicuous is the key to survival. Many reef dwellers begin life as minute larvae, designed like exquisite glass sculptures, and quite different from their adult forms.

Opposite: Magnificent scallops grow to over 10 cm in width. The blue spots on a scallop's mantle are primitive eyes that allow it to detect movement and light.

Life is not always subtle here. Sardines and anchovetas swarm in such vast numbers and densities that they seem to turn into a single being. The shoal looks cooperative, but in fact depends on each individual having a better chance of surviving within it than beyond it. A lone fish is easily picked off, whereas a dense shoal confuses and swamps the predators' reactions and appetite with sheer numbers. Or at least that's what the sardines hope. Sometimes the predators descend in such numbers themselves that even the largest shoal won't satiate them. Such hunted bait balls are the most remarkable spectacles of the open ocean.

Bait-ball feeding is a good time to meet the cast of creatures that otherwise live as elusive wanderers in the ocean – shadows across the blue. In the air, seabirds like masked boobies spear into the boiling sea and 'fly' underwater to propel themselves among the sharks, dolphins, sea lions, marlin, tuna and mackerel that are already there. Sometimes it seems that almost as soon as it began, the frenzy has dissipated and the predators dissolve away into the vastness. All that is left is a silver confetti of fish scales, glittering as they twist and spiral on their descent to the sea floor.

THE DEEP

In 1995, Fernandina split open and erupted molten lava into the sea on the west side of the volcano. The sea boiled so intensely that many fish were overwhelmed by the super-heating of water and floated to the surface. Among them were some deep-ocean fish never seen before and subsequently classified as new species. Amazingly this chance event was one of the only times scientists have collected deep-water life from Galápagos. The depths past 180 m are as much a mystery as the surface of the moon. Although they account for more area and potentially more species than the entire biomass of shallow habitat, access to their secrets is expensive and challenging to acquire.

The first known attempt to glean any knowledge of the area was the 1891 voyage of the US Fish Commission steamer *Albatross*. Its crew dragged nets and other collecting devices along the volcanic bottom and brought up a collection of fascinating creatures never seen before. Telescope-eye fish and the polka-dotted jello-nosefish were among many species new to science. Since then only three other expeditions have followed to document the species and habitats of this remote area, and of these one was as recent as 2006. Following in the footsteps of George Allan Hancock, the wealthy Californian who funded a number of marine-biological surveys in the 1930s, Paul Allen, co-founder of Microsoft, launched a deep-sea submersible mounted with a camera to photograph habitat beyond 300 m. As the submersible glided over the depths, the camera beamed back live images that revealed a fascinatingly beautiful world.

On these few expeditions almost every dive or bottom drag delivers a feast for the imagination and curiosity that holds remarkable parallels with the time when Charles Darwin visited Galápagos. Just as almost every walk and every cast of his net, hook and line recorded a creature new to the world, so every dive into the depths of the abyssal sea brings up species never seen before. It is a world that awaits discovery and enlightenment.

Opposite: Black-striped salema prefer the protection of the reef to the vulnerable open ocean. But living near land is no guarantee of safety with formidable predators like the Galápagos shark on the prowl.

7

By 1535, the year of the Galapagos islands' discovery, humans had already reached the shores of New Zealand, Rapa Nui, Hawaii and the Aleutian Islands. Thus, *the fate of the wildlife* in those remote places had been decided. The *mythical nature* of the Galápagos islands, with their fogs, currents and lack of water, provided some protection in the following centuries – *but it was not to last.*

CONSERVING THE GALÁPAGOS

Previous page: Puerto Villamil on Isabela has changed from a sleepy village of sandy roads and cautious people to an aggressively growing town with land prices soaring and illegal fishing rife.
Above: The Jessica *was wrecked in 2000 while supplying fuel for tour boats. It spilt over 900,000 litres of bunker and diesel. A heavy swell and the lack of a quick response made this a disaster.*

The abundance of giant tortoises to be found on Galápagos at the time Fray Tomás discovered them may have added to the stony appearance upon which he remarked. As we have seen, Tomás may well have been the first human to arrive here and it is intriguing that, like so many other organisms that have evolved to make these islands world renowned, he should have found himself there by accident.

By the time Darwin visited the islands in 1835 they were feeling the impact of human footsteps and suffering the effect of human hands, for they offered passing ships, including whalers, seal hunters and pirates, valued resources.

Still, they *were* rocky islands in the middle of nowhere. Yet they exert such a fascination that everyone wanted and wants to visit them; books are written and films are made about them. Why, for almost 200 years, have scientists investigated their reptiles, soils, birds, forests and oceans? And why, today, is everyone creating such a fuss about their conservation?

For the great thinkers of the nineteenth century (Malthus, Darwin and Wallace among them) who were questioning the nature of life, its survival, the growth of populations and life forms and the laws that controlled them, the issue of origin was of critical importance. From where did life come? And why, as Darwin noted, were the organisms different on two such physically similar volcanic archipelagos as the Cape Verde and Galápagos islands?

In the search for pristine environments where the prime factor of isolation had allowed organisms to evolve in unique ways, scientists were confronted by a brutal reality: that the human diaspora had reached the most remote places and destroyed life there. But in Galápagos humankind was still a stranger and the islands' wealth of remarkable species was undamaged by human hand. Lumbering reptiles in wild volcanic scenery excited people across the globe. Galápagos was prehistoric, a lost world for the imagination to play on and science to work on.

To this day, even with humans firmly planted here, Galápagos bestows this reality on the curious. But our presence is the reason there is need to talk about conservation, in concrete and active terms.

It seems as though at some time God had showered stones.

Fray Tomás de Berlanga, 1535

Galápagos is a household word and part of a world culture of mental challenge. It is an important place for science and human understanding. It is also a Mecca for human marvelling and enjoyment. It is a world of birds, of wild scenery, of active volcanoes, of unique animals, of mangrove inlets, and of marine mammals. World opinion (see box, page 164) clearly states that Galápagos is of inestimable value to world culture and that its unequalled environment must be spared destruction.

Yes, the word is out in Galápagos – conservation! But even given such important international recognition, how can the wonder of this island world be maintained? How can human ideas of development and lifestyle be integrated into its near pristine nature? What are the responsibilities of humankind – residents or visitors – in achieving this essential goal? Why are we at the centre of this debate?

Essentially we are a part of nature, our blood and bones bear witness to this fact, but the ability to manipulate resources and to accumulate profits from this manipulation is unique to our species. All other creatures survive, and are limited by, the availability of resources. If there is no water, plants will die. If there are no plants, the animals that feed on them will die too. In nature populations exist at a certain size because there is a balance between the availability of food and the number of consumers (not forgetting predators). This applies throughout nature and must extend to humans. Survival of native organisms in Galápagos is a day-to-day affair affected by many things, but including rainfall, climate change and capricious

INTERNATIONAL RECOGNITION OF GALÁPAGOS

The following time line gives an idea of how important this remote and rocky archipelago is in the eyes of the international community:

1936 First Ecuadorian laws specifically to protect Galápagos wildlife.

1959 Formation of the Galápagos National Park, which includes 97 per cent of the islands' land area. Creation of the Charles Darwin Foundation and Research Station as a result of alarm at the deteriorating ecological condition of the islands, especially in relation to the giant tortoises.

1978 At the request of Ecuador, Galápagos becomes a World Heritage Site under the auspices of the United Nations Educational, Scientific and Cultural Organization (UNESCO).

1984 Galápagos included in UNESCO's Man and the Biosphere programme, which attempts to develop bases for the sustainable use and conservation of biological diversity, and the improvement of the relationship between people and their environment globally.

1986 Galápagos Marine Resources Reserve established.

1990 Galápagos Whale Sanctuary created.

1998 Special Regime Law for the Conservation and Sustainable Development of the Galápagos Province passed by the Congress of the Ecuadorian Government. The Galápagos Marine Resources Reserve is defined and extended to 40 nautical miles outside the archipelago (from the original 15 specified in 1986).

2001 Galápagos Marine Resources Reserve included in the World Heritage Site.

2005 Galápagos declared an area of Specially Sensitive Waters by the International Maritime Organization (OMI).

events. At times, mass extinctions from natural causes have overwhelmed life on Earth. Today a mass extinction is occurring, but this time through our actions.

The fact is that, in a truly fundamental way, biological equilibrium has been altered all over the world, leading to the present disaster. Galápagos included. Pre-human Galápagos was a natural economy based on sunlight, the winds, the power of flight or fin, isolation, erupting volcanoes; it was a novel environment ready for exploitation by life. Post-human Galápagos is driven by human technology, the human invention of money, human desire and investment driving a very different economy.

Sustained economic growth, including growth of wages, takes no account of nature's ability to supply the basic resources for this growth, nor does it consider the way nature's reproductive power may vary from year to year. In Galápagos that variation can be drastic due to such events as El Niño, volcanic eruptions and forest fires. Yet the growth of a human-based economy, especially as a result of the dramatic rise in tourism, has drawn thousands of migrants. They demand more than the environment can supply and spend the resulting money in support of a lifestyle

Above: The fearless and endearing wildlife attract ever increasing numbers of visitors. The result is over-crowding and a growing risk of introducing exotic organisms.

that has little to do with hunter–gatherers. The sum effect is the degradation of the environment and the loss of the islands' geographical isolation, from the mainland and from each other. Isolation was, and is, the key factor in the origin of Galápagos wildlife. And today it is a vital factor in its protection.

It is not too sweeping a statement to say that human activity is responsible for all past and present conservation problems. As we have seen, the over-exploitation of resources began with the whales and tortoises and is rife to this day in the illegal fisheries for shark fins and sea cucumbers. The introduction of exotic plants – which now outnumber native ones – and animals has seriously threatened the unique flora and fauna of Galápagos. At the same time, changed land use and archipelago-wide maritime traffic, bringing with it among other things the contamination of water and land, disturbs fragile animal and plant populations and environments (see box, page 168). Tourism is guilty too: the ever increasing traffic to the islands puts additional demands on resources and contributes to the introduction of previously unknown organisms, including pests and diseases.

WATER AND THE CONSERVATION OF GALÁPAGOS

Scarcity of fresh water discouraged human colonization of the archipelago from the time it was discovered. Successful settlements started from the late nineteenth century, on islands were fresh water was found. Today, these islands – Floreana and San Cristóbal – are not the most developed and inhabited, but they bear witness to the length of human presence by the important extinction of species. Santa Cruz is the new centre of growth yet there is no fresh water there.

Permanent bubbling streams run down steep ravines and cascade off cliffs into the sea on San Cristóbal. A rather well-hidden treasure, which might have remained so had the impressive waterfalls not been so visible from the sea. Whaling vessels watered here, the *Beagle* did too. Closest to the continent by boat, this island became the capital, permitted the only intense farming effort to produce sugar for export, and supported a population of 1000 inhabitants in the mid-nineteenth century, while people on Santa Cruz lived off rain water and brackish water from *grietas* (crevasses in the lava).

Galápagos was still 'isolated' and water *was* a controlling factor to development. San Cristóbal supplied fresh water by barge to maintain the 3000-strong US-base human population on Baltra Island during the Second World War. And this, incidentally, sparked off a totally different Galápagos.

The Baltra airfield left by the Americans was the springboard from which tourism grew in the 1970s and made Santa Cruz the most accessible island. Development surged ahead. More water was pumped from *grietas* in Puerto Ayora. These became contaminated with salts and pathogens, degrading life on the islands that are expected to support populations that exceed supply. But development continued and desperate needs for new water sources along with it.

Nature has one answer – understand the supply and don't exceed it – but technology goes one better, providing desalination plants and diesel generators to run them. Brine is pumped into the sea. Smoke is pumped into the atmosphere, causing further contamination of both water and air.

Extinctions of reptiles, birds and mammals followed on San Cristóbal and Floreana, where a balance between man and nature was being respected. As technological advances bring new water supplies to arid lands, we are exponentially increasing the threats to wild plants and animals. Will more extinctions follow? Will human life suffer as well? Technologically, Galápagos could equal the Canary Islands with 1 million inhabitants and 10 million visitors. But at what ecological cost?

Previous page: Bartolomé – a unique and picturesque volcanic scene. On many days the anchorage below Pinacle Rock is congested with tour boats. Stairways have had to be constructed to prevent the erosion of volcanic dust.

The box on page 164 shows that the first effective conservation measures were adopted in the 1930s. The efforts that followed, particularly during the setting up of the Galápagos National Park in 1959, were initially successful in reducing pressure on the environment. As a backwater of Ecuador, and with very limited interest in the islands, there were no obstacles to this success. This downward trend in pressure finished about 1970, when tourism and large-scale migration began. The

Above: Damp, fern-festooned river-banks frame a waterfall on one of several permanent streams on San Cristóbal. A village of 6000 inhabitants depends on fresh water from one of them. Water and its treatment are a growing problem.

draw to the islands was initially the availability of work in tourism; later numbers were swelled by opportunists in the fishing sector, particularly once the lucrative sea-cucumber and shark-fin trades began to proliferate in the 1990s, fuelled by the rapid growth of wealth and demand for these commodities in south-east Asia. So will the overall pressure of human activities rise or fall in future years? This depends exclusively on the attitude of local residents and the love and care of all people who touch Galápagos soil. We control all the elements on which the 'pressure' depends. Ultimate success depends on human economic and social activities being ecologically acceptable. This initiative must come from the people themselves, since laws work by acceptance. This path is strewn with pitfalls. Many phenomena – including rapidly rising tourist numbers, the associated increase in air and maritime traffic and the external demand for resources – are directly opposed to maintaining the isolation of the islands.

TRAFFIC AND ITS DANGERS

Economic development has caused a huge increase in air and maritime traffic over the last 30 years. Today four cargo boats carry goods to the islands and 33 jets per

AN AUDIENCE WITH LONESOME GEORGE

Lonesome George is surely the most famous and melancholy of all the giant tortoises. He was found in 1971 on the goat-ravaged island of Pinta, already a fully adult male. In 1972 he was brought into enclosures in the Darwin Research Station because he seemed to be the very last tortoise on the island – the last wild member of the Pinta race. He has stayed there ever since.

Pinta was one of the relatively accessible tortoise islands that whalers exploited heavily in the eighteenth to early nineteenth century. When that was over goats arrived and reproduced prodigiously, leading George's race to final ruin.

When he was brought to captivity there was hope that a female Pinta tortoise could be found in the world's zoos. A $10,000 reward was offered, but no one took it. Attempts were made to mate George with tortoises from nearby races, but rumour has it that a tortoise's sexual organs atrophy with lack of use, and he had been alone a long time. He showed no interest in his blind dates. It has even been considered that George could be cloned. But with modern cryogenic freezing, he does not have to be alive to do that.

Today there is often quite a party atmosphere around Lonesome George's pen. Children and adults are delighted to meet the islands' foremost celebrity tortoise. His face adorns everything from posters to mugs and T-shirts. Jokes about Viagra abound. He is the patient, wrinkled face of conservation on these islands. Sometimes it is easy to forget that when Lonesome George dies, his entire race goes with him.

week bring passengers. Analogous to a landbridge, it exposes fragile and unique species to real and undeniable menaces.

Like most oceanic islands Galápagos used to be remarkably disease free. This is no longer the case, with human dengue, canine distemper, foot and mouth, swine fever and parvo virus all recorded. Small insects such as mosquitoes can create much more harm than a few bites. Two mosquito species have been introduced to Santa Cruz in the last 17 years. One of these can transmit human diseases such as dengue and yellow fever. The other is a carrier of three potentially fatal diseases for birds: avian malaria, avian pox and West Nile virus. Avian malaria has caused – and continues to cause – chaos among terrestrial endemic birds in Hawaii since its introduction 180 years ago. It is potentially a grave threat to Galápagos and it currently exists nearby on the South American continent. Avian pox has been in Galápagos for many years and is known to kill mockingbirds and Darwin's finches. It was probably brought there by infected, domestic chickens.

Above: A massive increase in fishermen and their boats has led to over-exploitation of marine resources. Attempts to set sustainable quotas has led to strikes and angry protests. Now the fishermen, and the marine wildlife they harvest, share an uncertain future.

West Nile virus is the new danger. It was brought to the New World in 1999 and now affects at least 280 New World bird species, including Magellanic penguins, close relatives of Galápagos penguins. The risks are frightening, for it is well known that organisms living in isolation have severely compromised immune systems and are susceptible to diseases to which continental organisms may have developed a strong immune response. There is no doubt at all that West Nile virus will reach Ecuador, considering its unstoppable southerly spread from the United States. Then it is only a matter of time before it arrives in Galápagos, borne by an infected mosquito within an aircraft or a larva in a pool of water on a boat. When it does arrive, the presence of the mosquito vector will make it impossible to contain the virus on an infected island. Tour boats or fishing vessels could transport the disease to other islands and should mosquitoes be present there it will spread.

To make matters worse Santa Cruz – where these mosquitoes already exist – maintains a wide selection of Darwin's finches, including the famous tool-using finch. The loss of these small passerines would be an ecological disaster, for they are the world's finest example of adaptive radiation, the process that demonstrates the ability of organisms to adapt to different environmental factors. Although all the known species of finch still exist within the archipelago, there have already been local extirpations (the extinction of a local population of a species) on several islands.

TURNING BACK TIME

Conservation of Galápagos will never be easy and putting the clock back to the time before human arrival is impossible. The giant tortoises, large-billed ground finches, barn owls, snakes and mockingbirds of Floreana will never be seen again. On the other hand, a thousand baby tortoises have been returned to Española, bred from 15 surviving adults after whalers had decimated the population and goats had severely affected the vegetation. As this chapter is being written, helicopters circle over the great shield volcanoes of Isabela, eliminating introduced herbivores, especially goats, whose insatiable appetites threaten whole ecosystems (see box below). The endemic vegetation, including the sunflower scalesia, is recovering on Santiago. Sperm whales swim in the protection of the whale sanctuary and fur seals scamper among the rocks. Restoring ancient ecosystems is a reality, refreshing and exciting!

PROJECT ISABELA

One of the most destructive human actions in Galápagos was the introduction of goats and pigs. These animals have created chaos in island ecosystems worldwide. Project Isabela is a bold programme to eradicate them from two of the Galápagos islands.

Ground hunting was carried out for decades on Santiago as a way of controlling pigs, which rooted up vegetation, dug up turtle nests and destroyed petrel burrows, and goats, which ruled the land by the thousands, destroying the native vegetation. Eradication there used to be a dream rather than a reality, even though the removal of goats had already been achieved on Española and Santa Fe – smaller, more arid islands of easier access. Santiago and other bigger islands needed new creative thought. This was brought into sharp focus during the 1990s, when an exploding population of goats threatened the survival of one of the largest populations of giant tortoises, on Alcedo volcano on Isabela.

Project Isabela, originally Project Alcedo, began its detailed planning stages in 1998, by bringing together the world's specialists on eradication techniques on a large scale. Two techniques were decided on: aerial hunting from helicopters and the use of 'Judas goats' – sterilized animals that bear radio collars and are placed deliberately on islands with feral goat populations. These animals can be located remotely by radio-direction finders. Planted animals eventually associate with and 'betray' wild goats because of their social nature, always tending to gang together. Hunters track the groups and remove the wild animals, leaving the Judas goats to carry on their work.

With a staff of over 50, including helicopter personnel, ground crew, hunters and support personnel, and at a cost of $5.2 million, Project Isabela has succeeded in eradicating both pigs (3000 animals) and goats (80,000) from Santiago, and the eradication of 60,000 goats on Isabela is in its final stages. The programme's remarkable results in restoring original ecosystems are a testimony to vision, careful planning and dedication to purpose. The conservation of Galápagos takes on new meaning.

Above: A cactus finch probes dew-covered, nectar-rich flowers. Darwin's finches may be threatened by the spread of diseases such as avian malaria and West Nile virus.

SOCIO-ECOSYSTEM MANAGEMENT

The Galápagos National Park Service of Ecuador and the International Charles Darwin Foundation cooperate to seek answers to very difficult problems, such as those of introduced plants described in CHAPTER 4 (see box, page 105), the vital need to understand marine resources, ecological relationships and the identification and analysis of new concerns such as the introduction of mosquitoes already mentioned. To deal with these issues, and many others that are essential for the well-being of the ecosystems, requires high-quality, non-political specialists and administrators backed by government and international organizations.

Laissez-faire will not do; proactivity and creativity must dominate. If there is to be real progress the decade and a half of politically motivated strikes and anger over the restricted use of natural resources must come to an end. There must be an open-minded will for all sectors to collaborate for the good of the place in which they work and live, and to be guided by a genuine concern for the future.

The idea of isolating nature from human nature is not currently proving viable; today it is giving way to the concept of integrating people into conservation goals. This is called socio-ecosystem management and it recognizes that humans are

LAND IGUANA CONSERVATION

A visit to the national park's tortoise-rearing facilities includes, as rather an afterthought, a few land iguanas. They occupy two pens down the wrong end of the track and some others hidden away for their own good. Once Darwin found land iguana holes on Santiago so numerous that it was impossible to camp there. Now they are gone.

The story of one race of land iguanas – on Baltra Island – is particularly remarkable. It includes an unlikely mix of the Second World War and some powerful guardian angels.

It began with a self-funded scientific expedition in 1932, when the multi-millionaire George Allan Hancock anchored near Baltra (then South Seymour). His team noticed that the iguanas were emaciated and saw no harm in moving some 72 of them to the iguana-less island of North Seymour, where things looked lusher. It was by way of an experiment, to see if they would survive and prosper. By today's standards it was highly irresponsible. On subsequent trips Hancock found the iguanas apparently thriving on both North and South Seymour, but then history took an unexpected turn.

With the advent of war the Americans needed a base to guard the Panama Canal from the Pacific side, and the unlovely little Galápagos island of South Seymour was their choice. They built a runway and a large air-force base, and all noticed the goats, cats and large land iguanas that surrounded them on 'the Rock'. It has been popularly assumed that what followed was four years of iguana target-practice by bored servicemen, resulting in their extinction on that island. But Galápagos expert John Woram has looked into the history and found a subtly different story.

It appears that when the servicemen got to the base, they were under the strictest orders not to kill any of the animals (even goats). Ex-servicemen swear this order was taken seriously, since the iguanas had influential friends: in 1944 President Roosevelt went on record saying, 'I would die happy if the state department could accomplish something,' that something being an international wildlife sanctuary on Galápagos. However, the servicemen also noticed that they never saw young iguanas, and this had been the experience of visitors to South Seymour as far back as 1923.

The construction of an army base and airport can't have helped, but it may have been the goats and feral cats already resident on the island that were the root cause of the iguanas' local extinction.

They would have killed off the younger animals, leaving just robust adults and resulting in what conservationists call a 'land of the living dead' – a population doomed through lack of recruitment. A higher rate of mortality among adults due to army vehicles and restricted ranges may have been all that was needed to polish them off on the island.

However, by moving some iguanas from South to North Seymour all those years before, Hancock had fortuitously allowed some breeding individuals to survive in exile. These were later bred at the Charles Darwin Research Station and their progeny returned to their ancestors' old haunts on what is now called Baltra.

Today, with the help of the national park, the population seems to be slowly increasing and you can once again see young iguanas on the Rock.

an important element in the system. However, including people means limiting numbers and making those people assume responsibility, not just make demands.

The Special Regime Law for Galápagos (1998) is a starting point in this process. It recognizes the Galápagos National Park and the Marine Resources Reserve, which it amplifies to 133,000 sq km, as protected areas. It regulates a forum in which the principal sectors – conservation and science, fishing and tourism – debate ideas and seek consensus on issues of use and the need to conserve marine resources. It tries to build bridges over troubled waters, including jealousy over tourist permits and the rational use of resources. It is a youthful law and a new concept. Dedication and vision will decide its success.

So, what is needed for the long-term survival of Galápagos?

First and most importantly, an island culture with conservation goals – and conservation pride – needs to be generated. This will be brought about primarily through education and awareness, which will in turn ensure respect for the resources that give the people of the islands their living.

To back this up there must be stable, efficient institutions that develop conservation programmes based on accepted limits on development, and carry them out. This must involve control of migration and prevent the introduction of plants, animals and diseases.

Then comes responsible tourism, which means limited, high-quality tourism rather than mass tourism, ensuring that all visitors are involved in such a way that they feel responsible for their actions. Information available to tourists should explain how things are done and why. This is real ecotourism and it must be serious.

All these need to be supported by guaranteed funding and a clear long-term policy for restoring and conserving the ecosystems, both terrestrial and marine. The government of Ecuador must define the destiny of Galápagos as a guiding principle that must never be bent to political whim.

The value of Galápagos as a refuge for the human mind and for science is inestimable. Therefore the concessions that we make as residents and visitors should equal our respect for this value. If no concessions are made in lifestyle and attitude, then the value of Galápagos reduces to near zero. We need to live in a way that has the least possible impact on the environment. Cold showers? That's a start! At least let the water be heated by the sun! Ecological boats offering only essential services to visitors who willingly reduce their demands in order to conserve the islands. Taking the time to live at the same rate as all the fabulous species. That is the real meaning of appreciation.

Should the sea lions on the beaches of San Cristóbal, who are very tolerant of humans, be ousted because humans are not tolerant of their voices or their communal life? Are housing estates compatible with the natural values of Galápagos? What about golf courses? Is introducing sport fishing consistent with non-extractive appreciation of nature, the basis of a national park? Is this not more of the hunter hunting? Is Galápagos being humanized to its detriment?

As demand increases, so do the ills of society – human disease, water contamination, violence, robbery. Everything and everyone loses by indifference. Is it possible to imagine a world in which people would willingly live in a self-sufficient way that would automatically limit the population? Would they willingly give up the possession of domestic dogs and cats, both a serious menace to wildlife? Can a culture of love for home dominate the human desire for wealth? Man's present gaze is gold tinted; nature's is fixed on survival.

Today the conservation of the islands has expanded its view from direct assistance to nature to include all the human issues that challenge its very survival. Holistic answers are needed to deal with wide-scale problems. These must include unappealing or unpopular issues, such as waste-water treatment and tourism quotas. Both the United Nations Development Program, through its Galapagos 20-20 plan, and the potential intervention of the World Bank, are multi-million dollar efforts to meet various challenges. The former concentrates on four major fields – tourism, fishing, migration and introduced species, and has over 60 projects in progress to achieve its goals. The latter plans to develop tourism on the mainland, as a means to diluting the tourism invasion to Galápagos but, in fact, will undoubtedly make Ecuador an even more popular place to visit and cause further problems on Galápagos. Institutional strengthening is considered by all to be vital but the need for institutional cooperation is just as important. Reinforcing the control of invasive species arriving on aircraft and boats, and maintaining observation of the Special Regime Law, needs multi-institutional support.

Thus, a massive international economic investment is being made to defend the Enchanted Isles. It is a challenge for humans to cure, or at least mitigate, a totally human-induced problem that threatens the survival of not only one of the most widely known archipelagos in the world, but an unsurpassed stronghold of inspiration where nature can be faced, raw and beautiful, and where we can come to grips with the power of evolution, the force of our own origins.

At a local and more scientific level, the need for international cooperation, for monitoring the human and natural environments, and for understanding the working of ecosystems, is being addressed by the Charles Darwin Foundation's new

Strategic Plan and the new Management Plan of the National Park. This is vital for resource management, but requires funding. Caring for tiny fragile populations of animals and plants in highly restricted habitats is also urgently required – Galápagos penguins and flightless cormorants being notable examples. Captive breeding is on the table to ensure the survival of mangrove finches. With introduced cats and rats surrounding and penetrating their unique habitat this may be the only way that the species, currently with fewer than 100 individuals, can survive. Rice rats and some plants may be candidates for similar programmes. The success of Project Isabela has sparked off new island-wide projects to eliminate introduced mammals such as goats. Eyes are now focusing on Floreana and San Cristóbal as potential sites for these operations in the near future.

The risk of losing the unique qualities of Galápagos is high. Both the National Park and the Marine Resources Reserve face grave problems. Part of the world community is now taking this as a serious matter, but what about the rest? Will the economic investment for conservation drive out the growth-related economic interests? Are we in the middle of a tug of war? The real force behind achievement is not based uniquely on economic investment but on human determination to see things work. This is not just at institutional level but throughout society.

Seeding and watering a Galápagos culture – a love of the land and a caring society, for itself and its environment – would resolve many of the human-related issues and pave the way to solving the more long-term problems. To achieve this, the Special Regime Law must be applied. Migration must be curtailed. Quality of life – a healthy mind in a healthy body – has to be achieved. Money is needed but philosophy and self help are also essential. Sometimes too much money can spoil humans and ruin good intentions, just like plants being given too much water. Will any amount of money solve the problems of the Galápagos if the real issue is the local attitude?

Galápagos can be spared the fate of so many other oceanic islands if we fulfil these conditions. The volcanoes of the archipelago explode when the pressure becomes too strong for the cooled surface to contain, pouring irresistible rivers of glowing magma over forests, into the sea where life dies ... Then time and life's resilience bring back the forests, renovate the sea. The ecosystems recover in their ancient ways. Humankind is a new and more deadly pressure, threatening permanent alteration to unique and irreplaceable biological systems. Yet we can consider the results of our actions, can choose to change ... it is for every individual to understand the nature and value of life and to demand ways that ensure its survival. The reward is not in taking but in giving.

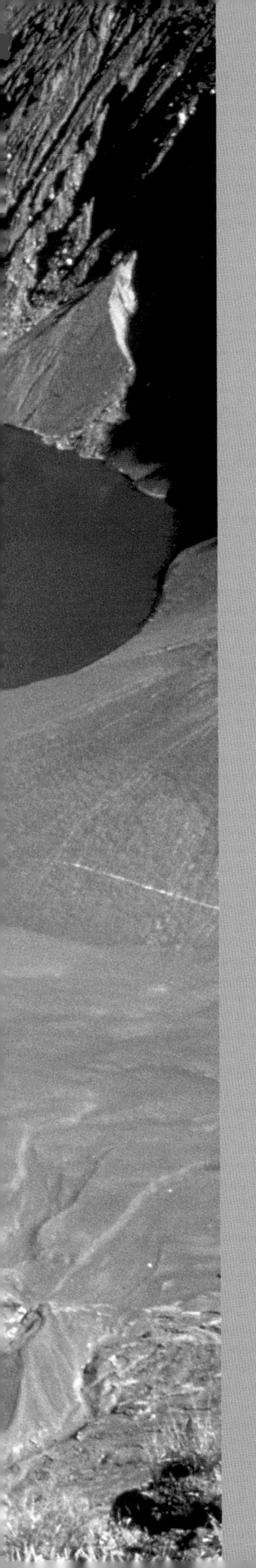

8

The Galápagos archipelago includes the most diverse, near pristine islands in the world. Their birds, animals and plants are *bizarre, conspicuously trusting* and *unique.* As such, the archipelago is of *inestimable value* to science and a source of *human inspiration* in a world of *disappearing wilderness* and increasing urbanization.

GALÁPAGOS – WORLD'S END

Rescued from over two centuries of destruction, the treasures of the Galápagos might seem safe, locked away as they are in a national park and the vast marine reserve that surrounds them. Yet in the last decades the islands have seen massive increases in their human population. There is still time left to decide whether these islands become just one more coastal fishing resort or fulfil a long-dreamed destiny as the place where humanity learns to live gently as the custodian of nature in a modern Eden.

Previous page: Looking down into the crater lake in Cerro Azul's remote caldera. The islands are in constant natural flux; this lake formed just 20 years ago. Above: Kicker Rock on San Cristóbal.

SCIENCE'S HALLOWED GROUND

Considering their still largely pristine nature, small size and defining isolation, it is remarkable that Galápagos is a household name; a group of islands that changed our world.

In the human realm, Galápagos has had its most profound impact on the natural sciences. Today many of us take for granted the fact that our ancestors were primates, that birds such as finches evolved from reptiles, and that we all developed from some very ancient replicating molecules. More than that, scientists understand that the mechanism for this change was natural selection. It is a revelation that has influenced everything from philosophy and culture to medicine and machine design. And it all started here on Galápagos with Darwin's visit less than 200 years ago.

Darwin's Galápagos insight was not limited to evolution. While collecting the islands' animals and observing the human impact upon them, he started to think about the dramatic ramifying effects of adding just one new predator to an island that was unaccustomed to 'the stranger's craft or power'. When Darwin wrote in *Origin of Species* that there is a 'web of complex relations' among species, moving 'onwards in ever increasing circles of complexity', he was laying the very foundation stones of ecology.

It would be wrong to think that Galápagos science stopped with Darwin – far from it. Hundreds of recent studies have confirmed and developed important biological theories here, and the work of Peter and Rosemary Grant on Darwin's finches (see CHAPTER 3) is the best demonstration we yet have of evolution actually occurring in our own time frame. And Galápagos science continues in groundbreaking vein. The fact that this is one of the most geologically active places on Earth has made it a new frontier in volcano science. Satellites have been trained on the islands and used to provide new insights on everything from the movements of continents to the giant volcanoes of Mars.

The islands had remained almost unchanged ... month after month, and year after year, on most of the islands the reptiles and birds and sealions knew only each other's forms, and alone watched the sun rise and set.

William Beebe,
Galapagos: World's End (1924)

More remarkable still have been the results of deep-sea submarines looking at the Galápagos rift 2 km down on the ocean bed. What they found was the twentieth century's biggest oceanographic discovery. Living at immense pressure around the super-heated, chemical-laden waters was a diverse, unknown and utterly unexpected community of creatures. Their lives were supported by metabolizing, energy-rich minerals ejected by deep-sea vents. Until then we had no idea life could even survive in such conditions, let alone that a whole biological community might thrive. That this world of extremes is supported by an energy source that has nothing to do with light from the sun caused a reassessment of our understanding of how life first evolved and where else in the universe we might find it. It was another insight into the origin of species, and there's still a lot of Galápagos abyss waiting to be explored.

GALÁPAGOS NOW

For all its scientific history and value, the main reason most people visit or treasure Galápagos is to experience its nature. There is still staggering variety and beauty here, as this book attests. What often touches people most during their time here is that it is also a wilderness where animals have never come to fear people. Our past sins seem to be absolved as a sea lion nibbles our flippers and calls us to play, a mockingbird unties our shoelaces or a young hawk lands on our backpack simply for a better view. If we had a childhood vision of what a passage through Eden would be like, this was it.

Much can be learned from the progress of human history on Galápagos. The simplest analysis of the past tells us that when natural resources are short, things get ugly. It's where we're all heading. But modern Galápagos found a non-subtractive way to use its wildlife via tourism and, for a while at least, it seemed that a local solution was in place to preserve the islands. Galápagos has an unrivalled history in the effort that has gone into its conservation. Ecuador takes deep national pride in its custodianship of one of the great natural wonders. Yet, as we saw in the previous chapter, today the islands are threatened and little by little are being destroyed.

OUT WITH THE OLD

As chance would have it, I write this as a tiny species of black thrip is moving across my keyboard trying to find its way back to the plants it is infesting outside my window. It is an insect that arrived on Galápagos this year. On the Galápagos of the last 3.5 million years I might have had to wait a thousand years or more to see a new arrival establish itself. But on the Galápagos of today new arrivals turn up many times each year, and many have the potential to devastate the island's endemic species and communities. As long as the islands host a rapidly growing human population supplied from the mainland, the problem will continue. Indeed there is a more general feeling here that the core concept of Galápagos as a preserve of nature is losing its way. The National Park and Marine Resources Reserve can pay for themselves and earn many a good living via regulated ecological tourism, but a gold rush free-for-all including commercial fishing and resort-style developments threatens an ancient natural treasure and a longer-term human future.

THE PARABLE OF THE FINCHES

There is a Galápagos parable from the Grants' work on Darwin's finches that author Jonathon Weiner relates. On the island of Daphne Major, a small volcano crumbling into the sea, the scientists found something rather disturbing. Many cactus flowers, on whose fruits the local finches depend, were sterilized by having their stigmas – the

Above: Daphne Major Island, scene of some remarkable recent discoveries about Darwin's finches.

part of the flower that receives pollen – neatly removed. Without the stigma, fruits cannot form, and without fruit the finches could starve later in the season. So who was sabotaging the cactus and the finches? Remarkably, it was the finches themselves.

Among the finch population a few individuals have learned that it's faster and a little more convenient to feed early at cactus flowers with the stigma removed. They pluck and snip by habit. They are a minority of the population, but they do a lot of damage. Of course all the finches stand to suffer from the resulting lack of fruit – but as individuals the flower vandals do slightly better than the average, and that's all that natural selection notices. They could mend their ways and suffer a slight disadvantage so that the whole population was better fed. But evolution involves no foresight and wouldn't reward them for it. The population as a whole might go into decline – but while it did the vandals would be doing the best. The finches on Daphne might even go extinct – and if they did the vandals would be among the last.

Human beings, of course, aren't very different from those finches. We peck and snip and see the world changing around us. We want to live a life above the average, and because of that the average falls. But we have foresight and the ability to reason, and that alone should allow us to do better. It is often asked, 'If we can't save Galápagos, then what can we save?' We might give ourselves a little hope for our own future if we could just manage to preserve these tiny, precious islands today.

PURE SHORES

On the furthest coast of Española, the oldest island at the southernmost tip of the archipelago, a vast curving bay of water-worn lava welcomes drifting fragments of the world beyond. The remains of that island's eradicated goats are here as well as the wreck of an early tour boat. This is not a visitors' site. Little rain falls, and no large animal now treads here beyond the broad padding feet of a waved albatross or the occasional researcher. As a consequence it is everywhere littered with the delicate paper-white bones of earlier generations. A walk here turns up lava ossuaries of whales, sea lions, finches, mockingbirds, tortoises and more. There are mummified corpses of female iguanas that died just as they were digging their nests. They were killed so that Galápagos hawks could bring food to their own young. Booby chicks fight life-and-death battles with their siblings, while parents gaze on impassively. Albatross fledglings soar from the cliffs to spend years on the ocean, or fail, to die as wreckage among the boulders. This whole island is at the end of its life, ultimately destined to sink and join the remnants of older islands in the deep ocean. But in its wake new islands are being born and colonized. It has been happening like this since before the birth of our own species, and it will continue long after we are gone.

If ever a place revealed how from Darwin's 'war of nature, from famine and death … endless forms most beautiful and most wonderful have been, and are being, evolved', then this is it. The sun and wind seem to breathe life into bleached bones as new generations replace the old. Darwin remarked that there was 'grandeur in this view of life' and indeed there is. From a high viewpoint on these cliffs at the furthest end of Española seabirds soar fearlessly at head height, while below them marine iguanas engage massive waves and frigatebirds mount their gaudy displays.

Beyond them to the north and west, on other Galápagos islands, ancient tortoises still walk in green glades that have never seen a human being, and land iguanas still risk life and limb to descend into the rumbling jaws of an active volcano. Sea lions surf for the joy of it and then pull iguanas' tails that the devil may care. Penguins turn silver shoals and killer whales round on giant manta rays. Flamingos march, giant cactuses bloom, mockingbirds sing, a finch prepares its tool, and a booby falls for the dapper blue feet of a suitor. Individuals from 6000 or more Galápagos species go about their daily business, honouring a four-billion-year-old pact with nature to survive and to reproduce. Their own world within our world, and a constant source of enlightenment, mystery and wonder.

Opposite: Punta Cevallos on Española, at the far southeastern edge of the archipelago. One of the oldest islands in Galápagos, it will soon – in geological terms – follow many other islands in the group by disappearing under the sea. Overleaf: Sunset over the Mariela Islet in Elizabeth Bay, off Isabela.

A trip to the Galápagos Islands is typically a *once-in-a-lifetime* experience. Though most tours and trips are well planned to show you the best of what the islands have to offer, with a bit of preparation of your own you can get even more out of the experience by making sure you choose a tour that covers what you most want to see. A number of excellent wildlife guides and historical and contemporary texts offer more in-depth information into specialist subjects, and a comprehensive list of these appears on pages 231–3. The following gazetteer gives an *essential background* to the *wildlife, geology* and *visitor* and *dive sites* of the islands, and is a *useful reference* before, during and even after your trip. It is intended to complement the main body of the book and to provide hints and suggestions to enable you to *enjoy Galápagos* in both a comfortable and conserving way.

GAZETTEER

LAND ANIMALS

Giant Tortoise (endemic)

Geochelone nigra (elephantopus)

The largest and best known of the Galápagos land fauna, the giant tortoises are slow-moving reptilian herbivores, browsing on a variety of herbage including grass, cactus and shrub. They are able to go several months or more without food or water, but drink voluminously and bathe when the opportunity arises. Adults of some dome-shelled subspecies may reach over 300 kg (one hit 400 kg in captivity) but less than half that is more typical. Ages of 150 years and upward are thought possible.

Previous page: Greater flamingos feed in a Floreana lagoon. Right: A dome-shelled giant tortoise from Santa Cruz.

During the mating season (around December–January) males engage in contests involving neck stretching and biting. Courtship often entails a chase. Mating is a noisy affair with loud, bellowing groans by the male, and 'ship's timber' creaking of the grinding carapaces. Females migrate to certain parts of the islands to lay their eggs in holes dug at traditional sites. All ages suffer from competition and habitat changes wrought by goats. Intensive conservation based around captive breeding and eradication of introduced species is allowing races threatened with extinction to stage something of a comeback, though it is unlikely tortoises will ever recover to their pre-contact numbers.

Best places to see: in the wild, in the Galapaguera of San Cristóbal and in the highlands of Santa Cruz. In the latter location, their appearance is seasonal – October, November, December are best, lessening in January, but increasing again after June. Volcano Alcedo was once a spectacular visitor site, but is closed at the time of writing. In captivity tortoises can be seen in the pens in the national-park stations on San Cristóbal highlands, Isabela (near Puerto Villamil), Floreana in the new highland tortoise-rearing facility, and on Santa Cruz (Puerto Ayora), where you can meet Lonesome George (see page 170).

Galápagos Land Iguana (endemic)

Conolophus subcristatus and *C. pallidus*

Found at scattered sites throughout the central and western parts of the archipelago. Santa Fe hosts its own species, *C. pallidus*, with a more drooping crest and redder eye. The large population of *C. subcristatus* on volcano Wolf includes mystery 'pink'-coloured iguanas whose taxonomic status is under review. Land iguanas eat mostly invertebrates when young, becoming vegetarian as adults. Adults can sometimes be seen climbing clumsily in search of forage. In lowland areas opuntia cactus flowers and pads are a key resource, allowing them to survive extreme drought for protracted periods. The land iguana's long, shallow tunnels provide insulation for cool nights and hot days.

Breeding behaviour has been observed on Fernandina. Males set up aggressively defended territories in June, which females enter for food or access to other sites. Males may guard small harems of females in the mating season, becoming very brightly coloured. There can be elaborate courtship rituals, including active female solicitation. Other males in less favourable territories

intercept females, running them down and biting the neck to hold on. On Fernandina pregnant females make a remarkable migration to the fumaroles on the crater rim (and even to within the caldera) to lay their eggs in July. Nest tunnels are guarded from other females with vigorous fighting, to ensure that eggs are not displaced or damaged. Young are very fast and cryptic, being vulnerable to many predators, including snakes and hawks.

Best places to see: South Plaza, Santa Fe, Cerro Dragon on Santa Cruz and Seymour Island. Watch out for them, too, near the kiosks of the airport at Baltra. They can be encountered throughout the year, and visited in captivity at the Darwin Station on Santa Cruz.

Lava Lizard (endemic)

Microlophus spp.

Small colourful lizards found throughout the major islands except Genovesa. There are seven recognized species, notable for their different sizes and colours. Males may reach 15–20 cm and females 12–18 cm; females are plainer, with red 'cheek' markings. They feed on plant material, small invertebrates and on occasion the young of their own species. Virtually every other predator on the island seems to eat them too, from centipede to hawk. A favourite lava-lizard behaviour is staking out sea lions or their corpses and catching the attracted flies with remarkably acrobatic leaps and lunges. Lava lizards can often be seen on promontories (including the heads of marine iguanas!), where they 'press-up display', watch the world and possibly cool off. They breed in the wet season, laying eggs in sandy burrows. Like the land and marine iguanas, females compete aggressively for these burrows, slapping each other with audible flicks of their tails.

Best places to see: in the lowlands and drier habitats.

Left: A land iguana in typical grooming pose. Below: A lava lizard with red 'cheeks' and an unpatterned torso – distinctively a female.

Galápagos Snakes (endemic)

Four endemic species of snake inhabit Galápagos. The Galápagos racer *Alsophis biseralis* is found in various subspecies around the southern and central islands, and the hood racer *Philodryas hoodensis* only on Española. The banded Galápagos snake *Antillophis slevini* occurs on Isabela, Fernandina and Pinzón, and the striped

Galápagos snake *Antillophis steindachneri* on Seymour, Baltra, Rábida, Santiago and a morph on Santa Cruz. They are very mildly venomous, unwilling to bite anything but their prey of small lizards and nestlings. The prey is subdued by constriction, the venom helping with subsequent digestion. When several of these snakes are seen together on beach lava or sand it is normally a sign that marine iguanas are hatching.

Best places to see: Punta Espinosa (on Fernandina) and Seymour.

Right: The Galápagos rice rat has a mousy appearance. Below: A Galápagos snake devouring a lava lizard.

Gecko (endemic)

Phyllodactylus spp.

There are six endemic species, each unique to a particular island. They feed primarily on insects. There are signs that the endemic species is being supplanted around the town of Puerto Ayora by the larger, introduced *P. reissi*, which can be differentiated from the native *P. galapagoensis* by its semi-translucent skin.

Best places to see: common around the lights in towns.

Rice Rat (endemic)

Oryzomys and *Nesoryzomys* spp.

Four species, all nocturnal, are found on the uninhabited islands. *N. fernandinae* and *N. narboroughii* are found on Fernandina, *N. swarthi* on Santiago and *O. bauri* on Santa Fe. They can be distinguished from the introduced black and Norwegian rats by their short tails, round ears and generally more 'mousy' appearance.

Best places to see: on Santa Fe towards dusk.

Darwin's Finches (endemic)

Thirteen species of blackish to brown-green birds, famous for some rather remarkable beaks, odd behaviour and a distinguished history in science.

Geospiza spp.

Very diverse genus with mostly black males and brownish-grey females, mainly herbivores and seed-eaters.

Large ground finch

G. magnirostris

The largest finch, immediately recognizable by its outsize deep 'nutcracker' bill. Tends to be solitary and is found in the lowlands and transition zone of the larger islands except Fernandina, San Cristóbal, Floreana and Española.

Medium ground finch
G. fortis
Medium-sized and very common, found throughout the archipelago in almost all habitats. Often seen around towns and also picking parasites from reptiles (which respond by adopting an upright stance for easy beak access).

Small ground finch
G. fuliginosa.
Very widespread with a small beak for eating small seeds. Another keen reptile hygienist.

Sharp-beaked ground finch
G. difficilis
Found only in the highlands of Santiago and Fernandina as well as Darwin and Wolf, where it picks off bird parasites and has also developed the vampire and egg-eating tendencies for which it is famous (see page 95).

Cactus ground finch
G. scandens
With a distinctive long beak, it specializes in nesting and feeding in opuntia – and often has a coating of pollen on its head. It is found throughout the islands wherever opuntia is available, with the exception of Española and Wolf.

Left: A cactus finch, with its characteristically long beak, inspects an opuntia.

Large cactus ground finch
G. conirostris
Fills in the gaps that *G. scandens* leaves, with subspecies at opposite ends of the archipelago on Genovesa and Española.

Camarhynchus spp.
Females tend to be greenish brown, with the males similar except for having a black head. Mostly insectivores.

Large tree finch
C. psittacula
Found mostly in the humid zone, this quite large species feeds primarily on invertebrates.

Medium tree finch
C. pauper
Smaller and restricted to the humid zone of Floreana.

Small tree finch
C. parvulus
Found in transitional and humid zones throughout the archipelago except Genovesa, Marchena, Española, Wolf and Darwin.

Cactospiza spp.
Two large, rather similar species with similar sexes.

Woodpecker finch
C. pallidus
This famous and noticeably hyperactive finch uses modified tools (twigs) to hunt for insects in holes. Widespread through the transition and dry zones of the larger islands excluding Floreana, Marchena and Española.

Mangrove finch
C. heliobates
The rarest of the finches, with fewer than 100 pairs in the coastal and mangrove forests of Isabela.

Vegetarian finch
Platyspiza crassirostris

Large and distinctive, the male with black head and back, extending to all parts of the body in some specimens. The female is olive-brown with a pale ochre belly and streaked breast. Found virtually throughout the archipelago in transitional and moist zones – though seldom abundantly – and eats leaves and shoots.

Warbler finch
Certhidea olivacea

The most petite finch, approximating to a warbler in both the shape of its beak and its melodious song.

Best places to see: the species of finch vary by island and habitat as noted above. To see a good variety of finches, keep an eye out around the tortoise-rearing pens at the Darwin Station .

Galápagos Hawk (endemic)
Buteo galapagoensis

An attractive and fearless raptor, the dominant land predator of Galápagos, found throughout the archipelago with the exception of San Cristóbal, Floreana and Santa Cruz, where people have hunted it out, and Genovesa, where its predatory mantle is adopted by the short-eared owl. It is thought that there are fewer than 1000 individual Galápagos hawks in the wild. The adult is some 55 cm high with a wingspan of over 1 m. Adults are dark brown; juveniles mottled brown and fawn. They hunt everything from centipedes and locusts to nesting marine iguanas, and also scavenge whenever the opportunity arises, eating sea-lion afterbirth and dead feral mammals.

Their social system is unusual in that one female can consort and indeed mate with up to four males, all of which may help rear the young. Nests are large affairs of sticks and bones, and there may be several in a single territory. They breed in arid areas and within volcanic calderas, but range effortlessly across the islands. Be careful approaching any occupied nest, and watch out behind you when any very young hawks land near by: their protective parents may not be far away.

Below: A dark adult Galápagos hawk.

Best places to see: at any time of year at Punta Espinosa on Fernandina, Punta Suarez on Española and Puerto Egas on Santiago.

Galápagos Mockingbird (endemic)
Nesomimus spp.

Inquisitive and opportunistic member of the arid-zone communities. There are four species, each inhabiting subtly different habitats. The Chatham mockingbird *N. melanotis* is found on San Cristóbal. The Charles mockingbird *N. trifasciatus* used to be resident of Floreana but is now found only on two of its offshore islets, Champion and Gardner. The hood mockingbird *N. macdonaldi* is the largest and most raptorial of the four, found only on Española. The most widespread is the Galápagos mockingbird *N. parvulus*, which has varieties and subspecies spread over the central, northerly and western islands.

The mockingbird's diet is varied, from fruits and nectar to insects and carrion. The tastes of the Española mockingbird, with its longer down-curved bill, are the most predatory. The birds cooperate to raise their chicks, with the parents being joined by the young of previous broods. Outside the breeding season groups of up to 20, including several breeding females, may defend territorial borders.

Best places to see: all main islands except Floreana. Española's Gardner Bay is where they often make the strongest impression.

Galápagos Dove (endemic)
Zenaida galapagoensis

Once very abundant in the dry zones, this bird was relentlessly hunted by people up to the late 1960s, and cats continue to take a heavy toll. Where such predators are absent it can still be quite common, as in the highlands of Española. Its main foods are seeds picked from the ground, small insects and more rarely cactus-fruit pulp. Courtship is a typical dove affair with aerial displays followed by low bows and cooing. The nest can be under rock overhangs or in opuntia cactus.

Best places to see: in drier habitats.

Short-eared Owl (endemic subspecies)
Asio flammeus galapagoensis

A mottled brown and pale brown owl with intensely yellow eyes offset by a dark face. Found throughout the archipelago except Fernandina and Wolf, and tends to be active in the day unless Galápagos hawks are found in the same area.

Best places to see: Genovesa, Santa Cruz near Media Luna, and around Puerto Villamil on Isabela.

Barn Owl (endemic subspecies)
Tyto alba punctatissima

A nocturnal–crepuscular hunter, remarkably curious and tolerant of people. Found throughout the islands with the exception of certain smaller barren or distant islets. It hunts for lizards and small mammals – but despite its good work removing agricultural pests, local people still superstitiously kill this bird and burn its lava-tunnel nests. It is claimed the owls are a major threat for chickens, but there is little proof of that. As a consequence they are becoming rarer on main islands such as Santa Cruz.

Below: Galápagos doves are reluctant flyers and are more usually seen walking.

Best places to see: quiet locations on most islands after dark. Listen for calls or subtle movement in trees.

Right: A female yellow warbler. Males have a red head cap.

Yellow Warbler (native)
Dendroica petechia aureola

A small, attractive and very visible bird throughout the islands' drier zones. Can sometimes be seen doing the rounds of sea-lion colonies and intertidal zones where it picks off the abundant small flies. In towns it often enters buildings to take sugar or engage in mock displays with mirrors. Its habit of foraging on roads leads to casualties on Santa Cruz where the melancholy sight of a bird attending a dead mate is all too common. Males can be distinguished from females by the red bar on the crown.

Best places to see: ubiquitous around the coasts of the islands.

Vermilion Flycatcher (native)
Pyrocephalus rubinus

Though not particularly numerous, this bird is memorably hard to miss or mistake. The vivid crimson of the adults is a photographer's dream, as is their habit of resting on giant tortoises. Though present in the drier zones, they are most often found in the moister transition and scalesia zones, making repeated flights from well-appointed perches to catch flying insects.

Best places to see: Isabela and Floreana.

Galápagos Rail (endemic)
Laterallus spilonotus

A diminutive, skulking bird of the damp highlands. Very unwilling to fly, it has been widely exterminated by predators as a consequence. Hard to see, but can be quite unafraid when located. It hunts for invertebrates on the ground, scraping the surface with its disproportionately large feet.

Best places to see: the miconia zones of Santa Cruz and San Cristóbal.

Smooth-billed Ani (introduced)
Crotophaga ani

Galápagos lore has it that these birds were introduced by farmers in the 1960s or 1970s to control ticks on cattle, but it seems they don't eat many ticks here. Farming has apparently provided suitable circumstances for a rapid boost in their numbers, now estimated in the thousands. They are also active well beyond the agricultural zone. It is possible that they compete aggressively with native insectivores and seed-eaters. They are noisy birds and can be seen in large groups of up to 40 individuals. Despite being weak fliers, they seem to be spreading among islands rather quickly.

Best places to see: the highlands of Santa Cruz and less commonly on Isabela, San Cristóbal, Floreana, Santiago, Santa Fe and Pinzón.

Green Hawkmoth (endemic subspecies)
Eumorpha labruscae yupanquii

The 15 species of hawkmoth are important nocturnal pollinators around the islands. The green hawkmoth's proboscis is approximately twice as long as its body, enough to probe the deepest of Galápagos flowers. The larvae feed on a highland vine, but the adults' rapid flight allows them to range across the islands.

Best places to see: flying rapidly around lights in towns.

Galápagos Blue Butterfly (endemic)
Leptodes parrhasioides

Tiny, iridescent and powder-blue butterfly, unusual among Galápagos butterflies because it is so delicate – many other species here, such as the monarch or milkweed butterfly *Danaus plexippus* and the sulphur butterfly *Phoebis sennae*, are of more rugged constitution, celebrated for their long-distance flights (the monarch fairly regularly makes it by accident across the Atlantic to Europe from the USA).

Best places to see: the dry zone, where the adult favours low flowers of acacia. Common after rain.

Galápagos Centipede (endemic)
Scolopendra galapagoensis

A fascinating endemic with an important predatory role in the arid ecosystems. Grows up to an impressive 30 cm, with large poison fangs capable of delivering a very painful bite to humans. Its size and weaponry allow it to take a range of food from other invertebrates up to lizards and nestlings. By day it generally stays out of harm's way in small burrows and crevices.

Best places to see: the dry zone at night. Increasingly rare around Puerto Ayora, which many ascribe to the impact of roads, cats and chickens.

Painted Locust (endemic)
Schistocerca melanocera

Large, colourful grasshopper. Immatures are green or brown depending on habitat and are taken by hawks, as well as lava lizards.

Best places to see: commonly found in the dry zones after rains.

Fire Ants (introduced)
Wasmania auropunctata (little red) and *Solenopsis geminata* (tropical)

Tiny 'aggressive' ants with a painful sting. The little red can be recognized by its slow movement. Both species seek sweet foods such as honeydew, and can commonly be found in thin, winding columns climbing the trunks of flowering cactus. They were accidentally introduced to the islands in the twentieth century and now occupy all the inhabited islands and many uninhabited ones. A serious threat to endemic invertebrates, they even imperil reptiles and nestlings due to their habit of swarming and stinging. Finding ways to eradicate these species is a conservation priority.

Best places to see: look carefully on cactus trunks around the Darwin Station for slowly moving little red fire ants. Larger tropical fire ants will probably bring themselves to your attention while walking in farmland areas on Santa Cruz.

Below: A painted locust, one of the more prominent insects after rain.

INLAND PLANTS

Scalesia – the sunflower tree (endemic)

One of the best examples of adaptive radiation on the islands. It is likely that the first ancestor of the group settled in the dry lowlands, and from there species evolved to fill a variety of niches around the islands. Some 15 species are known, of which the most familiar are:

Above: The epiphyte-rich under-storey of the tree scalesia forest. Right: A mature opuntia cactus.

Tree scalesia
S. pedunculata

Grows to the size of a tree and dominates moist zones in the highlands of Santa Cruz, San Cristóbal, Santiago and Floreana. Subtly different species of tree scalesia occupy similar niches on the various volcanoes of Isabela.

Radiate-headed scalesia
S. affinis

Looks much more like what you might imagine for a plant of the sunflower family: a long-stemmed shrub with pointed, somewhat furry leaves. The fuzz helps it hold on to moisture in the barren lava fields it colonizes. Found on Santa Cruz, Isabela and Floreana.

Heller's scalesia
S. helleri

A coastal plant with heavily divided crinkly leaves. Goats like eating it – so it has historically fared rather badly on Santa Cruz and Santa Fe, where it is found.

Putting the leaves of these three species together, you would find it hard to tell that they were related. Scalesias reveal family likenesses when they produce their delicate variants on a daisy-like flower.

Opuntia – giant prickly-pear cactus (endemic)

The different species of opuntia tend to divide themselves among the islands rather than among habitats, perhaps because their leaf structure and growth form is less plastic and their pre-adaptation to arid climates is more restricting than that of the scalesias.

O. echios

The largest species (up to 12 m) with various subspecies growing on the islands around Santa Cruz, including Santa Fe.

O. galapageia

Found on nearby Pinzón, Pinta, Rabida and Santiago, this species is smaller (to 5 m) with a noticeably conical trunk.

O. insularis

A long-spined opuntia that grows up to 5 m and is found only on Isabela and Fernandina. On volcano Cerro Azul it is found with *O. saxicola* and on Sierra Negra with *O. echios*.

O. helleri

The most unusual of the tree opuntias, found on Genovesa, Darwin, Wolf and Marchena. Often appears as a shambles of low-growing pads with little evidence of a trunk.

Candelabra Cactus (endemic)

Jasminocereus thouarsii

Distinctive giant cactus found throughout the arid zone; it can grow to 7 m tall. Flowers are cream to mauve-coloured and open before dawn. The large red fruits are edible but not very tasty.

Lava Cactus (endemic)

Brachycereus nesioticus

Found on bare lava, the orange-tipped stems are distinctive. Larger plant clumps (reaching 2 m in diameter) may well be more than 100 years old.

Palo Santo (native)

Bursera graveolens

Palo santo and its smaller cousin the dwarf palo santo *B. malacophylla* (from northern Santa Cruz and the associated islands) are common trees of the arid and transition zones. Their lichen-covered grey trunks and skeletal frame are very distinctive. They burst briefly into leaf after rain, producing white flowers and small, purplish fruits.

Parkinsonia (native)

Parkinsonia aculeata

Distinctive tree of the dry zone, 2–4 m in height. Numerous small 'stems' are modified from the mid-ribs of leaves and each leaf base holds a distinctive triple set of spines. The bark is green, allowing the whole plant to photosynthesize even when the leaves are lost.

Mollugo (endemic)

Mollugo flavescens

Low-growing species with yellowish leaves and white flowers, a pioneer on lava flows. There are five species, seen on the lava of Punta Espinosa and Bartolomé among others.

Muyuyu (native)

Cordia lutea

A familiar sight around towns and gardens with its vivid, green oval leaves and tight bunches of yellow trumpet flowers. Common in the arid and lower transition

Left: A candelabra cactus on its favoured habitat of rough lava. Below: The massed trumpet flowers of the muyuyu attract carpenter bees by day and hawkmoths at dusk.

zones. The fruit are translucent white, globular and very sticky.

Galápagos Cotton (endemic)
Gossypium barbadense

Distinctive shrub with large yellow flowers that open in the early morning, infuse red in the late afternoon and close at night. During the second day they fall. Seed capsules reveal a mass of fibrous, white lint, giving the plant its common name. Found throughout the main islands except Genovesa and Santiago. A close cousin, *G. klotzchianum*, is rarer with unlobed leaves and seeds without 'cotton'. It is found on Isabela, Floreana, San Cristóbal, Marchena and Santa Cruz.

Manzanillo, Poison Apple (native)
Hippomane mancinella

Large evergreen tree, often growing near underground sources of brackish water in drier zones. The fruit and sap are poisonous. Commonly seen on the inhabited islands, particularly due to its odd choice as an ornamental garden plant.

Galápagos Passion Flower (endemic subspecies)
Passiflora foetida var. galapagensis

Another beautiful Galápagos native (though its habit of low vine growth over rocky terrain makes it hard to appreciate); it is found on Floreana, Santa Cruz, Isabela and San Cristóbal. A far rarer endemic relative is *P. colinvauxii* from the moist zones of Santa Cruz. The latter is interesting because its leaves bear yellow-spot egg mimics, evolved in the genus to deter egg laying by heliconid butterflies (few of which ever reached Galápagos).

Miconia (endemic)
Miconia robinsoniana

Lending its name to the miconia zone near the moist island peaks, this plant is far less abundant now than in the past. Its oval leaves are heavily embossed by veins; the flowers are borne in small purple clusters. Found on San Cristóbal and Santa Cruz, but the remaining stands are gravely threatened by the intrusion of the far larger, introduced quinine tree.

Galápagos Tillandsia (endemic)
Tillandsia insularis

The only endemic species of the wide-ranging South American bromeliad flora. Its distinctive vase-shaped arrangement of leaves holds water, but not nearly so well as many of its relatives on the mainland. Finches often drink at these plants, and endemic land snails like to rest at the leaf bases. It is epiphytic and found in high transitional and moist zones of the main islands.

Ionopsis Orchid (native)
Ionopsis utricularioides

One of the 14 species of orchid found on Galápagos, an epiphyte of the transitional zone and moist uplands, with pretty but unshowy flowers. Also fairly common are the three species of *Hebenaria* ground orchids found on Santa Cruz, Isabela and Fernandina. They bear many pale flowers on upright spikes.

Lichens

Quite diverse on Galápagos: there are some 500 species and doubtless many to be discovered. The lowlands offer ample unshaded solid substrates, and enough airborne water in the form of garúa or moist sea breezes for them to thrive. The 'dyer's moss' *Roccella babingtonii* is actually a lichen, one of several species that festoon trees with long drapes. It has had an unusual role to play in the history of the islands: unsuccessful attempts to exploit it for dyes were the economic basis of some of the earliest failed colonies.

Marine Iguana (endemic)

Amblyrhynchus cristatus

The only marine-going lizard in the world, with only one species found in the archipelago but seven subspecies identified on different islands. Typically seen on land basking on the dark lava coasts to raise their body temperature sufficiently to allow it to enter the water to feed. They are vegetarians, most surviving on a strict diet of marine algae. During the breeding season (November–January), males are extremely territorial over small patches of shore or prominent rocks, while females fight for nesting grounds to lay their eggs in February and March.

Best places to see: Punta Espinosa (Fernandina) has good numbers throughout the year. Española is the best place to see the spectacular red and jade colouring of the males during breeding season. But look out for ever opportunistic basking iguanas anywhere there is a flat bit of coast in the sun.

Green Turtle (native)

Chelonia mydas agassisi

The species of turtle most commonly seen in Galápagos. Males are distinguishable from the females by being smaller with a

Left: Green turtles can often be encountered throughout the archipelago while diving and snorkelling. Below: Claws and strong limbs help the marine iguana negotiate rocky lava shores.

longer tail. From November to February groups of males can often be seen trying to mate with a single female in lagoons, beaches and shallow coastal areas. Throughout the year look out for heads bobbing up to breathe in open water, and turtles resting in mangroves and lagoon areas. Egg-laying occurs year round with a peak in December and January; females can be sometimes seen returning to the water in the early morning having laid their eggs in nests excavated high up the beach.

Best places to see: group mating in protected bays and lagoons, such as the beach at Punta Cormorant (Floreana), Tortuga Bay, Tortuga Negra (Santa Cruz) and Los Tunelos (Isabela). Also at sea when coming to surface to breathe. A regular cleaning station is found on the sandy bottom of Santa Fe bay.

Galápagos Penguin (endemic)

Spheniscus mendiculus

One of the smallest species in the world, standing only 35 cm tall, with a distinctive braying call. Galápagos penguins mate for life and are the only penguins to breed in the tropics. They form colonies, nesting in the shelter of lava cracks and crevices. Two eggs are normally laid, though typically only one chick survives.

Above: Galápagos is home to three-quarters of the world population of blue-footed boobies. Below: The characteristic Zorro-style mask and orange beak are distinguishing features of the Nazca booby.

Best places to see: onshore, at Punta Espinosa (Fernandina), Mariela Islets, Elizabeth Bay and Las Tintoreras (Isabela), and Bartolomé. At Bartolomé they can also be seen preening in the shallows near the beach in the morning or fishing under the shadow of Pinnacle Rock.

Flightless Cormorant (endemic)
Phalacrocorax harrisi
(formerly *Nannopterum harrisi*)

Stunning turquoise eyes and stubby wings make it unmistakable. The only cormorant on Galápagos, it fishes for eels, octopus and fish along the rocky shores of the western islands, swimming under water propelled by webbed feet. Its lovely courtship display often starts in the water with the pair circling, raising themselves out of the water and 'snake-necking'. It continues on land where mating takes place. The mounded nest, sizeable by Galápagos sea-bird standards, is constructed from the seaweed given mostly by the male as a part of the courtship ritual.

Best places to see: Elizabeth Bay and Urvina Bay (Isabela) and Punta Espinosa (Fernandina) offer chances to see cormorants on the nest, in the water and under water if snorkelling.

Blue-footed Booby (native)
Sula nebouxii excisa

The most commonly seen of the three resident booby species: they fish inshore waters and nest on the open ground in sizeable colonies at a number of key visitor sites. Their plunge-diving is an arresting spectacle and the distinguishing vivid blue feet can be seen even in flight. During the elaborate 'sky-pointing' courtship display, the male parades with a waddle of blue feet, arches his head, neck and back to bring tail, wing tips and beak pointing up, and lets out a whistle, while the female, stepping and sky-pointing, tucks her beak to her chest and honks.

Best places to see: keep at look out at sea throughout the archipelago. Punta Suarez (Española) and Seymour offer an unforgettable opportunity to walk among their colonies.

Nazca Booby (native)
Sula granti

The largest of three boobies, formally known as the masked booby, but classified as a new species in 2001. Colonies are typically found on cliff areas where updrafts and breeze make for easier take-offs. Adults are mostly white with black tail and primary feathers with distinctive, black face markings. Nazca boobies fish the waters between the islands

and so are seldom seen fishing. However, they can be seen cleaning or cooling off in the shallows upon leaving the nesting area. Courtship is similar to that of the blue-foot but not as elaborate and only the male exhibits the sky-pointing display.

Best places to see: Punta Suarez (Española), Darwin Bay (Genovesa).

Red-footed Booby (endemic subspecies)
Sula sula websteri

Nests in the outlying islands and fishes offshore, so although it is the most abundant of the three boobies, it is the least observed. It is also the lightest, able to nest in *Cryptocarpus*, mangroves, palo santo trees and other coastal vegetation. The bluish beak is often a key means of identification. Characteristic red prehensile feet and shorter legs enable it to perch on branches and even on ships' rigging. Two colour phases can be seen: a light-brown plumage, which is more common in Galápagos, and a white phase, occasionally leading to confusion with Nazca boobies. It is semi-nocturnal and typically exploits the offshore-fishing grounds associated with seamounts. It is even known to catch flying fish as they skim across the surface.

Best places to see: Darwin Bay (Genovesa), Punta Pitt (San Cristóbal), Wolf and Darwin, and rigging of boats.

Waved Albatross (endemic)
Phoebastria irrorata

Almost the entire world population of waved albatross is found on Española, with a handful of pairs on Isla de la Plata off the Ecuadorian coast. Española offers the take-off and landing runways required by this largest of Galápagos birds, weighing 4 kg with a wingspan stretching up to 2.5m. Wings, back and tail are light brown, with a white head and chest, and prominent eyebrows lend the bird a charming expression. Landings can be awkward and comical but their courtship is captivating. They face each other, tapping bills enthusiastically, until one breaks away, pointing its beak skywards. The other copies, they then coo, punctuated at intervals with snapping their beaks shut, then start the ritualistic joust once more. Mating occurs over five days, with eggs laid between April and June and hatching from mid-June to mid-August. The fledglings are brown balls of fluff – a far cry from the serene elegance of adults. Fledging takes 167 days during which the young undergo a variety of crazy hairstyles. Albatross may live 30–40 years and typically mate for life. They are nocturnal feeders and like squid.

Best places to see: on land at Punta Suarez (Española) between April and December and at sea throughout the archipelago.

Above: The waved albatross fishes across the Pacific and is the only species of albatross resident in the tropics.

Opposite: The distinctive wing patterns of the beautiful swallow-tailed gull are clearly seen in flight. When parent gulls return to the nest at night, the chick taps the white marking on the beak before feeding.

Galápagos Petrel (endemic)

Pterodroma phaeopygia

Identified by a black back, characteristic white forehead and underparts, and recognizable gliding and swerving flight. Nests in burrows and crevices in the ground of the moist uplands of the 'middle-aged' islands, returning to the same burrows each year. It fishes for small fish and squid and returns to the nesting areas under cover of night with particularly haunting calls, leaving before daybreak.

Best places to see: regularly seen at sea. Seen, but more commonly heard at night, in the highlands of Santa Cruz, San Cristóbal, Floreana, Santiago and Isabela, particularly during the breeding season.

Galápagos Shearwater (endemic)

Puffinus subalaris

Similar in appearance to the Galápagos petrel, but smaller with a black head and back, and white underneath and throat. Widespread across the archipelago, typically seen near noddies and pelicans, feeding from or just below the sea's surface where they can dive to a number of metres in pursuit of fish, crustaceans and squid. They breed on smaller islets, nesting in crevices and holes in the cliffs.

Best places to see: Plaza, Santa Cruz, Genovesa, South Plazas and the open ocean.

Madeiran Storm Petrel (native)

Oceanodroma castro

The storm petrels are the smallest of the seabirds found in Galápagos. Three species breed here and all are dark in colour with a white rump and characteristic petrel tubular nostrils. These are the largest, distinguished by a conspicuous white band on the tail, hence their other name of band-rumped storm petrel. Like the other two species they feed on small fish and squid from the sea surface, but are active both day and night. They nest in the shelter of lava crevices and ledges.

Best places to see: North and South Plaza, Daphne Mayor, Cowley, Devil's Crown (Floreana), Isla Pitt and Genovesa.

Galápagos Storm Petrel (endemic)

Oceanodroma tethys tethys

Distinguished by the white, wedge-shaped mark at the base of the tail, hence its alternative name of wedge-rumped storm petrel. Of the three petrel species, this is the bird most commonly seen at its nesting grounds by day.

Best places to see: Genovesa, Punta Pitt (San Cristóbal) and Roca Redonda.

Elliot's Storm Petrel (endemic subspecies)

Oceanodroma gracilis galapagoensis

The most commonly encountered of the petrels, also known as the white-vented storm petrel, often seen feeding from the sea's surface and around boats at anchor. Distinguished by the white lens-shaped mark on its tail, its small size and the feet that extend back beyond its tail while in flight.

Best places to see: from boats at sea throughout the islands. No recorded nesting site of this species to date.

Red-billed Tropicbird (native)

Phaethonaethereus mesonauta

A particularly elegant bird, graced with a brilliant white plumage and two long tail feathers. A black line runs across the eyes and a characteristic red beak makes it unmistakable. Also known as bosun birds, they fish offshore during the day, plunge-diving to feed on small fish and squid. They are typically seen returning to their cliff nesting grounds, and the protection of ledges and rocks.

Best places to see: colonies on Plaza, Seymour, Daphne, Rábida and Floreana.

Swallow-tailed Gull (endemic)
Creagrus (Larus) furcatus

An exquisite-looking gull with a distinctive red eye ring, black head and wing tips, pale grey back with white underside and bright red feet. The world's only nocturnal gull, it feeds offshore at night and in doing so avoids competition with other coastal species. It can be seen on the wing around dusk, or during the day at its nesting sites in cliff areas. It possesses a rich set of distinctive calls, including a piercing scream and an unusual rattle. Some are known to nest in Colombia.

Best places to see: colonies on Plaza, Seymour, Genovesa and Española.

Lava Gull (endemic)
Larus fuliginosus

Aptly named as its dark-grey colour and black beak and feet blend well with the lava coast. One of the rarest gulls in the world, with perhaps just 400 pairs, but can be seen throughout the archipelago. Unlike most coastal birds in Galápagos, lava gulls make solitary nests; the adults guard the blue-green coloured eggs fiercely, often with a shrieking call.

Best places to see: South Plaza, Puerto Ayora (Santa Cruz), Villamil (Isabela) and Genovesa. Can be spotted feeding on crabs, debris and waste on the seashore, or perched on the rails of a tour boat looking hopeful.

Brown Noddy (endemic subspecies)
Anous stolidus galapagensis

Despite the common name, the body can range in colour from brown to a sooty grey and fades to white on the top of the head. Brown noddies are opportunistic breeders and nest in the shelter of cliffs throughout the archipelago. It is from their nodding courtship that the common name is derived. They are inshore feeders, typically fishing without settling on the water, and are often encountered with feeding Galápagos shearwaters, blue-footed boobies and pelicans.

Right: The pouch of the male frigatebird serves as a major attraction for suitable females. Below: Brown pelicans are often seen flying in groups. They appear surprisingly elegant in flight for such a large bird.

Best places to see: colonies on Isabela, Santa Cruz and Seymour.

Brown Pelican (endemic subspecies)
Pelecanus occidentalis urinator

A real crowd pleaser, easily recognizable, whether perched on land or soaring in flight. A large grey-brown bird with a white head, webbed feet and characteristic long beak. Its pouch enlarges when plunge-diving, ballooning out to scoop up litres of sea water containing small fish and crustaceans, which are then drained and swallowed. Pelicans nest in mangrove or salt bush. Two or three eggs are laid, and the young look particularly gawky!

Best places to see: nest on most islands but more commonly encountered around boats and docks, particularly when a catch is being cleaned.

Magnificent Frigatebird (endemic subspecies)
Fregata magnificens magnificens

The larger of the two species of frigatebird resident in Galápagos, males are black with a purplish sheen and during the mating season develop a distinctive, engorged, red throat pouch. Females are black with a white breast. Curiously, frigatebirds have a small preen gland and are unable to waterproof their plumage effectively, so must avoid settling on the water. Magnificent frigates fish and scavenge the inshore waters and coast, feeding on fish, crustaceans, turtle hatchlings and even storm petrels. They are also well known for forcing boobies returning from fishing to regurgitate their catch, which they then intercept. They nest in trees and bushes bordering beaches, often using materials stolen from neighbouring nests. Courtship is a mesmerizing affair with the male positioning himself in a nest and using his inflated red pouch and distinctive warbling call to attract a female.

Best places to see: North Seymour, Genovesa and Kicker Rock (San Cristóbal).

Great Frigatebird (native)
Fregata minor ridgwayi

The male can only really be distinguished from his magnificent cousin by a greenish tinge on the back of what is otherwise black plumage. The female great is easier to identify due to the distinctive red eye ring and white plumage extending further up the throat to the chin. The great frigate feeds further offshore than the magnificent but will also mob boobies, forcing regurgitation. The two frigatebird species nest in the same areas and courtship is similar.

Best places to see: colonies on Daphne, North Seymour, Genovesa, Seymour and Kicker Rock.

Great Blue Heron (native)
Ardea herodias

Greyish blue across the wings and back, with paler chest and neck. The largest heron in Galápagos, it nests throughout the year, typically in mangroves, and feeds on small fish, turtle hatchlings and young iguanas.

Best places to see: lagoons and beaches throughout the islands.

Lava Heron (endemic)
Butorides sundevalli

A distinctive, small, grey heron with silver legs that turn orange during the breeding season. Closely related to *B. striata* and may not yet deserve species status. Commonly seen poised on mangrove roots waiting to strike at small fish or crabs, or picking its way along the shore and tidal areas, hunting sally lightfoot crabs, lizards and insects.

Best places to see: on the lava coasts and among the mangroves throughout the islands.

Night Heron (endemic subspecies)
Nyctanassa violacea pauper

Medium-sized heron, most easily identified by the black head, with yellow and white feathers extending behind the head and a white stripe beneath the eye. Most active at night, feeding on locusts, crabs, scorpions and other insects, often beneath street lights and in intertidal areas.

Best places to see: in the mangroves or in towns throughout the archipelago.

Cattle Egret (introduced)
Bubulcus ibis

Small, all white heron, with yellow legs in adults and darker leg colours in juveniles. This recent arrival to the islands is considered introduced because its establishment directly follows the introduction of cattle. It nests in mangroves and can often be seen in the dawn or dusk flying from coastal nesting grounds to agricultural land.

Best places to see: on shore or in agricultural and arid zones feeding on lizards and insects, often near livestock or grazing tortoises.

Common or Great Egret (native)
Ardea alba

Largest of the pure white herons, distinguished from the cattle egret by its size and black legs in the adult. Feeds from a similar selection of lizards, insects, small fish and crustaceans to the great blue heron. It too nests in mangroves.

Best places to see: lagoons and coastal areas of the central islands.

Greater Flamingo (endemic subspecies)
Phoenicopterus ruber glyphorhynchus

The striking pink colour and unmistakable curved neck and long legs stand out from all other coastal birds in the archipelago. Found

Right: White-cheeked pintails are usually found in groups in sheltered ponds and lagoons across the archipelago.

throughout the islands in salt-water lagoons that provide their food of shrimp and small crustaceans. The population typically moves among a selection of suitable sites, depending on conditions. Nests are small mud mounds, where a single egg is laid. Easily disturbed, care must be taken observing them, especially during the breeding season.

Best places to see: Punta Cormorant (Floreana), Las Bachas and Garapatera (Santa Cruz), Bainbridge Rocks and town lagoon of Villamil (Isabela), and Espumilla Beach (Santiago).

Common Stilt (native)
Himantopus himantopus

A lagoon bird with striking black and white coloration: black back and head, white chest and neck and distinctive white spots surrounding eyes and forehead. Distinctive long, red legs and feet.

Best places to see: feeding on small invertebrates and fish in the lagoons and intertidal areas throughout the islands.

White-Cheeked Pintail (endemic subspecies)
Anas bahamensis galapagensis

Characterized by white cheeks and a red dash on the side of the bill, otherwise speckled brown. Typically seen feeding on the surface of ponds and lagoons with nests protected in the surrounding vegetation.

Best places to see: sheltered lagoons and ponds throughout the islands.

Whimbrel (migrant)
Numenius phaeopus

The largest of the wading birds encountered on the shoreline, a migrant that is present year-round and is easily distinguished by its large size, long, downward-curving bill and distinctive shrieking call.

Best places to see: throughout the beaches and lagoons of the islands.

Wandering Tattler (migrant)
Heteroscelus incanus

Light grey migrant with characteristic barred

plumage underneath. Commonly seen on the beaches and in the lagoons. Has a distinctive call.

Best places to see: common throughout the coasts of the archipelago, usually searching the rocky shore.

Sanderling (migrant)
Crocethia alba

Most often observed scurrying back and forth along the waterline with the waves. Can be identified by the grey back with a white belly and throat, dark primary feathers and medium-sized black beak.

Best places to see: throughout the beaches of the islands.

Ruddy Turnstone (migrant)
Arenaria interpres

Distinctive chocolate-brown and black back with white underneath, a black collar around throat and orange legs. The slightly up-curved beak is used to turn over shells and stones while feeding along the coast.

Best places to see: widespread on beaches throughout the islands.

Galápagos Sea Lion (endemic)
Zalophus wollebaeki

Arguably the most charismatic resident of Galápagos and a tourist's favourite. Seen throughout the islands, particularly on beaches where they form colonies. Males are distinguished by their larger size (up to 250 kg compared to 120 kg for females) and humped forehead. Males are strongly territorial during the mating season (May to January). Birth on land is often a noisy and awkward affair for new mothers. Pups gather in groups, in protected shallows or rock pools. Males without territories form bachelor colonies on stretches of less hospitable coast.

Best places to see: widespread throughout the islands. Colonies on beaches of South Plaza, Santa Fe, Española, San Cristóbal harbour and Seymour. Look out for sea lions surfing the breaks at Punta Suarez (Española), Seymour and off Puerto Baquerizo Moreno (San Cristóbal).

Left: Galápagos sea lions are often encountered while diving or snorkelling throughout the islands.

Right: The thick pelt of the Galápagos fur seal, which helps guard against the cool water, almost caused its demise. Hunting up until the early 1900s greatly reduced its numbers but the population has since recovered.
Below: Sally lightfoot crabs are predated by some species of fish, herons and other coastal birds. When threatened, they let out a squirt of water, possibly as a means of defence.

Galápagos Fur Seal (endemic)

Arctocephalus galapagoensis

In the same family (*Otariidae*) as sea lions, fur seals are distinguished by their smaller size (males typically up to 75 kg, females up to 35 kg), flatter face and bulbous eyes. They are more agile climbers and so are commonly seen in higher or more awkward snoozing places on the rocky shore. They prefer rocky coast that offers deeper water immediately offshore, and shaded ledges and grottoes to escape the heat of the day. They feed at night on squid and fish, hence the big eyes. A thick pelt guards them against the cool waters where fishing is richest.

Best places to see: associated with cooler waters and rocky shores around Fernandina, Puerto Egas (Santiago), Marchena, Pinta, Wolf and Darwin.

Sally Lightfoot Crab (native)

Grapsus grapsus

Unmistakable crab. Adults are brilliantly red and orange with pale-blue undersides and juveniles much darker. The common name derives from their ability to skip across pools of water as they move through the rocky intertidal zone where they scavenge and hunt. They feed on algae, detritus, other crustaceans and even their own species, and can be seen picking dead skin and parasites from marine iguanas.

Best places to see: common on rocky shores throughout the islands.

Ghost Crab (native)

Ocypode gaudichaudii

Seen on the sandy intertidal during low tide when they emerge from burrows to search the beach for detritus and organic material. Look out for the burrows and balls of sand left as a result of their feeding.

Best places to see: common on beaches with fine sand throughout the islands.

Galápagos Hermit Crab (endemic)

Calcinus explorator

Like all hermit crabs, this endemic adopts mollusc shells for shelter and protection and can be distinguished from other species such as the semi-terrestrial land hermit crab (*Coenobita compressus*) by the dark-brown colour of its body.

Best places to see: typically on the rocky intertidal throughout the islands.

Ulva

Species of the bright green algae *Ulva*, also known as sea lettuce, form the staple diet of marine iguanas. Also fed on by green turtles, urchins, sally lightfoot crabs and a host of fish. This and the foliose red algae are the most abundant and easily identifiable algae species found on the rocky shore throughout the archipelago.

Best places to see: widespread across the archipelago.

Red Mangrove (native)
Rhizophora mangle

Commonest of the four species of mangrove found in Galápagos; it provides a home and habitat for many marine creatures and nesting coastal birds. Found in or near the water's edge in sheltered beaches and lagoons throughout the islands, distinguished by thick dark-green leaves and a network of prop roots. Flowers have four petals and the long fruits or seedlings drop into the water to be carried away by currents. These seedlings or, more correctly, propagules are particularly hardy and often found washed up on beaches across the islands.

Best places to see: Puerto Ayora, Tortuga Bay, Elizabeth Bay (Isabela), Punta Espinosa (Fernandina) and Darwin Bay (Genovesa).

Black Mangrove (native)
Avicennia germinans

Tallest-growing of the mangroves found throughout the islands, reaching to a height of 25 m. The dark colour of the bark lends it its common name. It colonizes areas of shallow, intertidal mud and sand with roots that help stabilize the substrate, coping with the waterlogged conditions by forming breathing roots (pneumatophores) that can be seen sticking out of the mud or sand.

Best places to see: Puerto Ayora, Tortuga Bay, Las Bachas (Santa Cruz), Elizabeth Bay (Isabela), Fernandina and Rábida.

Left: The prop roots of the red mangrove stabilize sifting sands – the first step in creating new habitat in the coastal zone.

White Mangrove (native)

Laguncularia racemosa

Not common but can be seen alongside red mangrove in lagoons. Distinguished by its paler, oval leaves, greenish-white flowers with five petals and pale green fruits.

Best places to see: Isabela, Santa Cruz and Floreana.

Button Mangrove (native)

Conocarpus erectus

The least common of the Galápagos mangroves, found on sandy beaches or near lagoons. Key identifying characteristics are the smaller, leathery leaves and balls of flowers that develop into the brown round fruits from which it derives its common name.

Best places to see: Puerto Ayora and Villamil (Isabela).

Salt Bush (native)

Cryptocarpus pyriformis

Common plant seen on many beaches across the archipelago. With thick, waxy leaves and small green flowers, the thickets of this low, evergreen shrub provide nesting for frigatebirds and other coastal species.

Best places to see: borders the road on the walk to the Charles Darwin Research Institute and National Park. Other sites: Punta Suarez and Gardner Bay (Española), Darwin Bay (Genovesa), Punta Cormorant and Post Office Bay (Floreana), Santa Fe and Bartolomé.

Galápagos Carpetweed (native)

Sesuvium edmonstonei

This endemic forms carpets of fleshy cylindrical leaves at the top of beaches. A perennial herb with a rich orange-red colour during the dry season, turning green and producing small, white, star-shaped flowers following the rains. The common carpetweed (*S. portulacastrum*) is similar but distinguished by pink flowers and flatter leaves.

Best places to see: South Plaza, North Seymour and Punta Suarez (Española).

Galápagos Purslane (endemic)

Portulaca howellii

Found on lava in the coastal and arid zones, this endemic is characterized by low-growing mats with fleshy stems and yellow flowers that blossom following the rains. It is now more commonly seen on offshore islands, where goats are absent.

Best places to see: Plazas, Santa Fe, Seymour.

Beach Morning Glory (native)

Ipomoea pes-caprae

Networks of vines up to 10 m long of this perennial can be seen helping to stabilize sand in beaches across the archipelago. The goat-foot-like lobes of its elliptical leaf give it its species name. It has a distinctive purple, trumpet-shaped flower and dark-brown hairy seeds that are resistant to sea water and aid the plant's wide dispersal.

Best places to see: beaches of Isabela, Santa Cruz and Genovesa.

Ink Berry or Sea Grape (native)

Scaevola plumieri

Keep an eye open for this shrub growing to 1 m in height on or close to beaches. It has waxy, rounded leaves and long, white flowers, and is easily identifiable when in fruit by the dark grape-like seed that is tolerant to salt water and aids in its worldwide dispersal.

Best places to see: Santa Cruz, Floreana, San Cristóbal and Isabela.

Blue Whale (native)

The blue whale is the largest mammal, perhaps the largest animal, ever to exist on the Earth. Its tall (over 6 m high), straight blow is often the best way to identify one from a distance. The blue whale is long, up to 30 m, and streamlined, with the head making up nearly one-fourth of its total body length. Blue-grey in colour, often with a mottled appearance, they can weigh over 100 tonnes, with some females weighing as much as 150 tonnes.

The blue whale is thought to feed almost exclusively on krill, up to 4 tonnes a day. They are most common around the western side of the archipelago, around Fernandina and Isabela, where they migrate during the winter months. They are fast, strong swimmers, capable of reaching 50 kph when they are disturbed.

Best places to see: around the western archipelago. Normally seen alone or as a pair; very rarely in larger pods.

Orca (native)

Orcinus orca

Also known as the killer whale, this is the largest of the dolphin family, growing up to 9.6 m and weighing up to 4.5 tonnes. Males have a more pronounced dorsal fin than females. A formidable hunter commanding the pinnacle of the food chain. In Galápagos it feeds off other whales, dolphins, sea lions, rays, sharks, mola mola, tuna and a host of other species.

Best places to see: around the archipelago in pods ranging from a few individuals to larger groups of ten or more.

Bottle-nose Dolphin (native)

Tursiops truncates

The most common dolphin species in Galápagos, often seen bow-waving boats and leaping out of the water in large groups. Highly social animals with complex group dynamics. They feed on small fish, eels and squid by using echolocation.

Best places to see: sometimes seen by divers near the shallow coastal fringes but mainly found in the open ocean.

Left: No one knows if the orcas found in Galápagos are residents or migratory. Their position at the top of the food chain is never better illustrated than when they are seen hunting sperm whales in Galápagos.

Whale Shark (native)

Rhincodon typus

The largest shark in the ocean, 14–18 m long. Feeds by filtering out fish and crustaceans, so it is thought to be common in the western archipelago where plankton is abundant.

Best places to see: a common sight off the most northern island of Darwin and to a lesser extent Wolf. Has been seen less frequently around the central and southern islands.

Scalloped Hammerhead Shark (native)

Sphyrna lewini

At birth, hammerheads are about 42–55 cm; adults can reach 4.2 m. During the day they school, occasionally in groups of 500 or more. At night they swim in smaller groups or as solitary individuals looking for food. Their wedge-shaped heads look like a primitive design, as if they were the precursors to the more refined design we associate with most sharks. In fact the opposite is true. The heads are a modern improvement in hydrodynamics and sense perception. Eyes located on the sides of the head give them good peripheral vision. It is thought that their spade-like shape gives them added lift, which compensates for their lack of a swim bladder. Barbed stingray tail bones are very occasionally embedded around the underparts of their jaws as evidence of their risky dietary preferences. They also feed on other fishes, cephalopods and crustaceans.

Hammerheads are often found in massive numbers converging off seamounts or peripheral open-ocean islets. Studies indicate that they swim in patterns and it's thought they use electromagnetic fields in the Earth's crust to navigate to and from these seamounts.

Best places to see: off Wolf and Darwin, also off Roca Redonda, Cape Marshall and Marchena. Can be spotted near the central islands of Gordon Rocks and North Seymour, although heavy illegal fishing has depleted their numbers in the central islands.

Right: Although scalloped hammerhead sharks can appear intimidating to novice divers and snorkellers, they are not considered dangerous sharks.

Galápagos Shark or Grey Reef Whaler (native)

Carcharhinus galapagensis

Maximum length 3.7 m, maximum weight 86 kg. Sometimes form large aggregations but are often seen individually. Believed to feed mainly off fishes, but also take squid, octopus and possibly sea lions. They approach without much caution and are inquisitive. This is a common tropical shark found offshore near or on insular or continental shelves. It is seen regularly near deep drop-offs near islands and is capable of crossing considerable distances of open ocean between islands.

Best places to see: near deep drop-offs on oceanic islets such as Enderby (Floreana), Kicker Rock (San Cristóbal), Roca Redonda, Wolf and Darwin.

Pacific Manta Ray (native)

Manta hamiltoni

The manta, sometimes known as the devil ray, because of the horn-like fin extending from its head, is the largest ray in the Galápagos. They measure 5–7 m across and weigh up to 1.3 tonnes. They feed by unfurling the cephalic fins, using them to funnel plankton, little fish and crustaceans into their mouths. Juveniles hide from predators under sand until they are old enough to migrate into the water column where they live as adults. Can often be seen leaping dramatically out of the ocean.

Best places to see: off Cape Marshall all year around but also widespread around the archipelago, especially off Gordon Rocks, Cousins Rock and North Seymour.

Spotted Eagle Ray (native)

Aetobatus narinari

Grows to at least 2.3 m wide and 1.2 m long. Often found near sand flats where they burrow for clams, conches, octopus, snails, marine worms and oysters. They have very powerful jaws and flat, chevron-shaped teeth for crushing shells. If their agility and speed fail to protect them from predators, their final deterrent is a long, barbed tail that is capable of inflicting a painful blow.

Above: An eagle ray looking for prey under the sand, which is where most of their food comes from.

Best places to see: the Galápagos is one of the best places in the world to see these shy creatures and divers and snorkellers can get very close on occasion. Widespread, especially near sand flats that are their main feeding grounds.

Moorish Idol (native)

Zanclus canescens

Colourful tropical fish from the western Pacific, common in warm waters especially around northern islands, where they can school in large numbers around coral reefs. They feed off algae, small invertebrates, especially sponges, and sometimes plankton.

Best places to see: in the warmer waters of the northern islands, especially around Wolf and Darwin. Can occasionally be seen in aggregations of over 40 fish.

Above: Batfish move over the sea floor sporadically, using their tail fin followed by a hop-like walk using their pectoral wings.

Red-lipped Batfish (native)
Ogcocephalus darwini

Batfish are found mainly on sand bottoms. They often take up the colour of the substrate, indicating an advantage in remaining hidden. They feed on snails, small molluscs and crabs.

Best places to see: Punta Cormorant (Floreana), Rábida, James Bay (Santiago) and Tagus Cove (Isabela).

Steel Pampano (native)
Trachinotus stilbe

Usually found in schools just under the surface, where they feed on plankton. Form large sphere-like schools, perhaps for breeding. When chased by sea lions they form tight aggregations, probably to confuse predators.

Best places to see: everywhere in Galápagos – Roca Redonda, Gordon Rocks and Enderby.

Long-nose Hawkfish (native)
Oxycirrhites typus

Derives its name from the way it perches on the coral awaiting the opportunity to feed 'like a hawk'. Found on walls, near endemic black coral and gorgonians, and rocky reefs. It feeds on tiny fish and larvae.

Best places to see: everywhere in the archipelago, but particularly Cousins Rock (Santiago), Champion Islet (Floreana) and Punta Vicente Roca (Isabela).

Stone Scorpion Fish or Pacific Spotted Scorpion Fish (native)
Scorpaena plumieri mystes

The only scorpion fish in the eastern Pacific. It relies on camouflage against the rocky substrate; here it will rest motionless until it sees a small fish that it will bolt up to and grab. For its own defence it has poison-tipped spines that it raises when threatened.

Best places to see: archipelago-wide but hard to see. Divers should be especially wary before placing hands on the sea bed.

Yellow-tail Surgeonfish (native)
Prionurus laticlavius

Derives its name from three white protrusions running in a horizontal line on the caudal peduncle that look like the tips of a surgical scalpel. Feeds on algae in large schools and is often attacked by various species of the territorial damselfish. The young are sometimes found swimming in the shallows very close to the surge zone and in tide pools.

Best places to see: one of the most ubiquitous fishes in shallow waters around the archipelago.

Chestnut Moray or Panamic Green Moray (native)
Gymnothorax castaneus

Common and one of the largest of the moray eels, averaging 1.8 m in length, but can grow up to 2.5 m and weigh up to 29 kg. The dark-green-to-brown colour comes from a yellowish mucus that covers the bluish skin and provides a defence against parasites and infectious bacteria. Found along rocky shorelines, reefs and mangroves, often camouflaged to hide in the reef from unsuspecting prey. They have very poor eyesight and use a keen sense of smell to hunt, feeding mainly at night. They are also very sensitive to the vibrations of wounded fish and will vigorously hunt down the victim. Depth ranges from 1 to 30 m. These eels can be territorial and have been known to occupy a specific reef for many years. The large mouth features strong, pointed teeth. The body is muscular and scaleless with a long dorsal fin that extends down its length, starting from the head and ending in a short tail fin.

Best places to see: off all the islands, but prolific numbers off Wolf and Darwin. Sometimes seen swimming during the day – unusual behaviour for this supposedly nocturnal creature.

Garden Eel (native)
Heteroconger klausewitzi
Endemic eel found in large aggregations inhabiting the sand flats and feeding on plankton as it drifts past their burrows. They dwell in holes that they also use to hide in when threatened. They prefer areas where the currents are stronger and the plankton is more abundant. At night they remain deep in their burrows.

Best places to see: where sand flats are predominant, mostly off the north and central islands. Less common in the west.

Commercial Sea Cucumber (native)
Stichopus fuscus
More easily seen than most other sea cucumbers, with a large, tube-like body with thick brown sides and soft spines. Heavily fished, so its numbers have been significantly reduced. Known as the worm of the sea, it recycles nutrients and re-oxygenates the substrate.

Best places to see: most abundant in the western archipelago. Often found hiding under ledges or rocks.

Green Sea Urchin (endemic)
Lytechinus semituberculatus
Small endemic, common in the rocky subtidal substrate and also found in intertidal pools. Occasionally covers itself with debris. Can sometimes be seen infesting the algae beds off which other creatures feed.

Best places to see: widespread; non-divers will even find them in intertidal pools.

Blue-striped Sea Slug (endemic)
Tambja mullineri
Small endemic blue-black nudibranch, common in the substrate, although it vanishes with El Niños. The conspicuous appendages near the head are known as rhinophores; the protrusions on the back are the gills. Found on rocky substrate from the intertidal zone down to depths of 45 m. A favourite food of *Roboastra*, another type of nudibranch that follows the chemical trail it leaves.

Best places to see: widespread; particularly off North Seymour, at 10–20 m. Easy to spot if divers look hard enough on the substrate.

Cup Coral (native)
Tubastrea coccinea
The most colourful of all corals in the Galápagos – bright yellow to orange and pink. They mostly feed at night, extending their tentacles to capture plankton.

Best places to see: widespread; most prolific in shallow water on walls, under ledges or in caves.

WILDLIFE CALENDAR

Galápagos can be visited throughout the year, as good wildlife watching is assured. However, seasonal variations in climate affect wildlife activity and should be considered when planning a trip.

Galápagos has two main seasons: the wet or hot season from about January to June and the dry from July to December, though the onset of the seasons varies from year to year and the transition months of December–January and May–June can bring a mixed bag of clear days followed by cloud and mist. The wet season is characterized by warm temperatures (26–32°C) and occasional heavy rainfall. The principal rainfall occurs between January and March, and the first falls stimulate a greening up on land; a Galápagos 'springtime'. April through to May offers an opportunity to see good land activity in predictably fine weather and the chance to see albatross on Española.

The dry season, known locally as the garúa, is characterized by cooler temperatures (22–24°C) with grey skies of high, blanketing cloud. The influence of the Humboldt current is at its strongest around September and the cooler water this brings marks an increase in plankton production, which drives a rise in productivity through the marine food chain. The seas are rougher from August to November.

The following is a selection of highlights of the Galápagos wildlife year.

Wet Season: January–June

Giant Tortoises

Breeding coincides with the wet season and can be best seen at El Chato and other private lands open to the public on Santa Cruz.

Marine Iguanas

Punta Suarez, Española, during December–February is the place to see male marine iguanas in their vivid jade green and red mating colours. Watch out for territorial scuffles between males across the rocky shore and later, from January to March, between females vying over choice nesting grounds in the lava sand.

Land Iguanas

Look out for nesting females during the wet season on the Plazas but during the dry season on Santa Fe. They may also be seen feasting on the portulaca succulent's blooms during the wet season.

Lava Lizards

Mating takes place at the start of the wet season, with even more vigorous displays of energetic head-bobbing than usual and territorial disputes between members of the same sex.

Finches

The first heavy rains and the glut of insects, larvae, seeds and berries that they bring stimulate the breeding time for the finches. Look out for nest building in opuntia begun by the male to entice a mate.

Frigatebirds

Seeing the huge, inflated, red gular sacks accompanied with a warbling call of a displaying male frigatebird is a highlight of many people's trip. The flamboyant courtship of the great frigatebird is best seen on Darwin Bay (Genovesa) between March and April, but magnificent frigatebirds can be witnessed year round on North Seymour.

Bats

During early evening in Puerto Ayora the endemic species *Lasiurus brachyotis* can be seen hunting at the level of the street lights, while the hoary bat *L. cinereus* hunts out of sight higher above.

Insect Explosion

The rains trigger an explosion in numbers of insects. Carpenter bees, the islands' most important pollinator, are seen busily visiting yellow flowers or investigating holes bored in dry wood. Keep an eye open for painted locusts, a variety of moths and some of the eight species of butterfly seen throughout the vegetation zones.

Spring Flowers

After the first rains (January to April) the flora of the arid zone erupts with new life. Palo santo, the endemic cacti *Jasminocereus*, *Brachycereus* and opuntia and the exquisite Galápagos passion flower burst into flower. From January to March the usually deep-red ground-covering succulent *Sesuvium*, produces mats of vivid pink (*S. portulacastrum*) and white (*S. edmostonei*) flowers.

Whales

Although distributed throughout the archipelago, most whale sightings occur in the rich, Cromwell current-fed waters off western Isabela and Fernandina. February is a key time to encounter sperm whales and even the great blue whales can occasionally be seen. Throughout the year keep an eye open for orcas, Bryde's whales, minke whales and dolphins, especially when travelling through the Bolivar Channel.

Spawning Aggregations

Broadcast spawning can be seen in Creole fish gathering in tight schools in dive sites such as Cousins (see page 229) during February and March. Look out for females bursting upwards, releasing their eggs into the water column, followed immediately by a trailing group of males who release their sperm in the hope of fertilization.

Green Turtles

Egg laying may occur throughout the year but is most common between January and June, usually timed to coincide with a high spring tide occurring during the night.

Dry Season: July–December

Galápagos Sea Lions and Fur Seals

The boisterous sea-lion mating season runs between May and December with a peak in September to November, though it can vary by a matter of months between islands. Look out for bulls defending their territories and seeing off rivals in frenzied torpedo chases in the shallows. Mating coincides with the peak of pupping as sea lions pup on 12-month cycle, so you may also see newborns. Fur-seal breeding shadows that of the sea lion but is less commonly witnessed.

Waved Albatross

These magnificent birds are present on Española only from April until the end of December, having left the colony to fish far out at sea during the wet season. Fluffball chicks and later gawky-looking fledglings can be seen at nests on Punta Suarez from June through to December and the captivating courtship display can be seen throughout the time the birds are on land but it is most common towards the end of the breeding season.

Galápagos Penguins

Breeding can occur throughout the year but is most common in September.

Boobies

Blue-footed colonies breed opportunistically when good fishing allows but usually between May and December. Their enthralling mating dance can be seen on Punta Suarez on Española, North Seymour or Genovesa. Nazca boobies can be seen displaying and breeding nearly year round on Genovesa and on Española between November and February.

Flightless Cormorants

These birds breed opportunistically throughout the year but particularly between November and March.

Green Turtles

Aggregations of mating green turtles can be seen throughout the islands in lagoons and off beaches during December to January.

Manta Rays

Cape Marshall is a great stretch of coast to encounter manta rays. These filter feeders are seen in largest numbers in September through to December during the sea's most productive period.

GEOLOGY

Though understandably famed for its wildlife, Galápagos is also a fantastic place to see geology. The forces and processes involved in an eruption are superbly preserved here, while gaseous fumaroles and the aftermath of recent activity help you to appreciate that Galápagos really does sit on one of the most volcanically active regions on Earth.

Lava Flows

The fluidity of Galápagos lava emitted during an eruption means lava flows are often vast and initially fast flowing. As the lava becomes more viscous and slows, the surface of the flow cools in contact with air, forming a crust that typically solidifies in two ways.

The fantastic, intestinal patterns of **pahoehoe lava** occur when the thin crust stops flowing while the molten lava beneath continues, pushing the surface into rope-like buckles and folds. Sullivan Bay on Santiago is an excellent place to see this. Look out for the outlines of trees that were engulfed by the flow.

Rough, broken **aa lava** forms when volcanic debris, boulders and the broken surface of the flow collectively build on the surface as the flow cools. This type of lava is found throughout the islands, but a particularly good example can be seen at the end of the Punta Espinosa trail (Fernandina). It is sharp and unforgiving, difficult for both humans and wildlife to walk over. At the Perry isthmus at the waist of Isabela, aa flows have created an unsurpassable barrier for giant tortoises, separating the populations found on the Alcedo and Sierra Negra volcanoes.

While on these barren 'new' lava flows, keep an eye open for sparsely scattered lichens and plants, such as the lava cactus *Brachycereus*, pioneering the first footholds on this new landscape.

Cavernous **lava tubes** are formed when the surface of a lava flow cools but the insulated core continues to flow quickly, leaving an open tube or tunnel when eventually the flow finishes. The most accessible examples are found on Santa Cruz, on private land near Bellavista and El Chato. They reveal the scale of lava flows, and in places reach 10 m high.

Where series of tubes and lava flows meet the coast, a network of caves, passages and grottoes can form. At the fur-seal grottoes on James Bay, Santiago, the broken surface of the lava offers a view of the sheltered pools used by fur seals during the day to escape from the equatorial sun. Nearby is a feature known as 'Darwin's toilet' – a small pool that appears to flush.

Los Tunelos, near Villamil on Isabela, is an alternative site to see the broken network of semi-submerged tubes. Exploring this rabbit warren of waterways is best done by kayak.

At **Los Gemelos** (the Twins) on Santa Cruz, the collapsed roof of what were once magma chambers has produced two deep pit craters. In the sheer sides of the craters the strata of different lava flows can be seen.

'Hornitos' or 'driblet cones' are small chimneys of lava, formed when pockets of gas permeate the molten-lava flow, throwing up small amounts of lava that land on nearby crust. Good examples can be seen on Sullivan Bay.

Spatter cones are also produced by gas escaping from the vent site, though on a bigger scale. Escaping gas throws up molten material that builds up in layers, creating the cones. If it happens that no lava flows in an eruption, the shape of these cones is preserved, as seen on Bartolomé.

When large amounts of rock material (tephra) are ejected during an eruption, they fall back to earth, creating large **scoria or cinder cones**, commonly seen throughout the islands, with particularly good examples inland on Floreana.

Tuff cones, also referred to as ash rings, occur near the coast or offshore. Contact with water creates a more explosive force and more fragmented tephra, so these cones are made up largely of ash. The finer particles, reacting with the heat of the eruption or a later process of cementing, lead to a particularly dense and harder rock known as tuff. Daphne Major and Minor are prime examples of tuff cones.

Sulphurous gaseous emissions are clear indicators of the pressure brewing beneath the surface of the islands; where they break the surface **fumaroles** are formed. The inner slopes of Sierra Negra and Alcedo volcanoes on Isabela are key places to see this. Underwater fumaroles can be seen while diving at Roca Redonda, one of a handful of places in the world where such volcanic activity can be observed comfortably within the limits of scuba diving.

The aftermath of a recent eruption can be witnessed on Isabela. Sierra Negra, close to Villamil, erupted in October 2005 and a trip to its rim and vents helps gauge the scale of the event. Over four days, lava flowed in rivers covering the eastern floor of the crater. Further around the rim towards the vents, the covering of scoria reveals the extent of fall-out from the explosion that produced a plume 20-km high.

Beaches

Colour and composition of the beaches vary across the archipelago. White beaches consist predominantly of the remains of seashells and corals pulverized over eons by the strength of the sea. They are typically found in the path of the prevailing southeasterly winds and waves that collect this material. Tortuga Bay (Santa Cruz) and Gardner Bay (Española) are beautiful examples. Black beaches originate from scoria fall-out following eruptions; good examples are Black Beach on Floreana and Fernandina. Brown and red beaches result from erosion of basalt rock and can be found on Rábida. On beaches in Floreana, such as Punta Cormorant, eroded cliffs containing olivine produce bejewelled beaches tinged with this green semi-precious stone and pieces of quartz-like feldspar.

Sea cliffs formed by erosion of the lava substrate can reach many tens of metres high. Punta Suarez on Española and Prince Philip's Steps on Genovesa offer good views of such cliffs. The blow hole at Española is best viewed at high tide.

WHAT TO EXPECT/WHAT TO TAKE

To appreciate the wonders of Galápagos fully it is highly recommended to take a live-aboard boat tour much as Charles Darwin did in 1835. Many of the most spectacular sites are too far away to visit in a day boat from land-based hotels. If you plan to visit in the peak seasons from December to March and July to August, you should book well in advance. Most tour operators conduct boat maintenance in September and October, so fewer boats are available at this time.

From the boat you will generally be escorted on two excursions per day, lasting an average of two hours each and rarely exceeding three. You may be slightly disgruntled that your tours are so synchronized and well planned that you have little time to yourself to explore and enjoy the place, but by law a guide has to accompany you on almost all the trails. This fact alone limits the worry of dealing with challenging situations, whether that means falling on uneven lava trails or encounters with territorial sea lions.

Inflatable pangas or zodiacs will take you from your boat to the trail head or beach from which excursions start, and you will be given a life jacket to wear until you are safely on shore. Sometimes the panga will pull up alongside a jetty; at others you have to paddle through shallow water. Your guide will warn you in advance whether to expect a 'wet' or 'dry' landing. Rubber sandals for stepping into the shallow wave zones of beaches will help you avoid minor discomfort on particularly abrasive coralline beaches.

Even for dry landings it is prudent to pack any delicate items in a waterproof bag. Landings are usually very safe, but occasionally wave action can cause the pangas to shift and people have been known to make unexpectedly wet landings!

Most trails are easy to walk on, but a few special wildlife hotspots are more challenging to access. For these places ankle boots and walking sticks are advisable. Your guide will warn you which trails these are. A commonly overlooked accessory is a hand towel for rubbing sand off your feet after a beach landing.

The equatorial sun can be very harsh, so wide-brimmed hats are recommended. Water, sun cream, lip block, mosquito repellent and a small first-aid pack are all useful items to carry, ideally in a backpack. Before each excursion cover any exposed skin with sun cream – don't under-estimate the damage sun can cause on cloudy days.

There is very little opportunity for shopping when on a live-aboard tour. Most boats are equipped with adequate supplies of medicine but tend to lack toiletries. It is better to veer on the side of caution and bring basic supplies of such things as shampoo, toothpaste, tampons, facial tissues, high-SPF sun block and lip block. Seasickness medication, sunburn-relief ointment, antibiotic cream, aspirin, plasters for blisters, ammonia solution, antibiotic pills and ear and eye drops all deserve a place in your first-aid kit.

Equipment for snorkelling

Most tour operators will provide snorkelling gear, but you might like to bring a well-fitting mask that will not leak and a pair of comfortable fins. A 3-mm wetsuit is adequate for snorkelling (and many people do without). If you can't get enough of those sea lions or penguins a 5-mm suit would be more appropriate in the cold season.

Equipment for diving

It is prudent to take a full set of your own diving equipment, even if you intend to dive only occasionally. It is your life-support system and familiarity with your gear reduces the risk of an accident.

The most common mistake divers make is in assuming that the ocean on the equator is always warm. Galápagos has cool waters for a tropical destination. Outside El Niño years, a full 7-mm one-piece wetsuit, preferably with a hooded vest, is recommended. If you are diving in the cold season for an extended period of time, you will be more comfortable with two layers of 7 mm in any combination, plus hood and gloves. Weights and tanks are provided by operators. A reputable dive operation will supply safety equipment, either an inflatable tube or a low-pressure dive alarm connected to your low-pressure hose, to use for alerting surface support. If you want to be extra safe you can invest in an inflatable tube attached to a dive reel and cord that you deploy while doing your safety stop at 5 m. This allows surface support more time to spot you before you come up. Do not dive in Galápagos without at least one of these devices. Currents can be strong and have been known to sweep divers out to sea.

If you do not have your own equipment you can rent it from dive stores. Be sure to try on wetsuits before you rent them and insist on connecting your regulator and BC to the cylinder to check the equipment is in perfect working order.

PHOTOGRAPHY

Photographing Galápagos wildlife is a real pleasure. Its celebrated tameness allows for careful composition and intimate portraits. The choice of camera is a personal one, be it video, film or digital. Whatever you use, make sure you have enough media (tapes, film or cards). A disposable, waterproof camera is a nice piece of kit for the non-specialist under water. If you plan on taking stills with an SLR camera, equip yourself with a UV filter, polarizing filter, wide-angle zoom and a moderate zoom lens. A telephoto lens is a great asset even on Galápagos, so consider 200 mm or even longer to get those surfing sea lions or plunge-diving boobies.

A major factor in good photography is the quality of light. Generally sunlight on Galápagos is best for photography early and late. Up to nine in the morning is optimal, past ten it can be too harsh, with shadows and highlights beyond the range of film or digital to deal with. In the afternoon, after about three o'clock is best. The contrast of dark lava with clouds or sea foam can be extreme, and a graduated filter can help keep the foreground brighter in wide photos.

Conditions on Galápagos can be hard on equipment. Take a splash-proof or waterproof padded bag to keep the camera safe during landings, and include a cleaning kit with a brush and a blower for sand and dust. A daily once-over with a brush and slightly damp cloth will stop dirt migrating into the works of the camera. While taking photos on the coast you may not always be aware of the fine spray blowing from the sea, so check your lens frequently. Though not recommended long-term, breathing on the lens to condense water and then wiping with a specialist lens tissue removes the haze of salt residue (which alcohol tends to smear). Removing suntan lotion from a lens is hard work, but lens-cleaner solution will help.

Animals on Galápagos are tame, but not without common sense, so keep your approach slow and on the trail: if the animal is reacting to you, you have gone too close. Learn how to turn off your flash, and always ask your guide before using it near animals. Angling in light from a small pop-up reflector can often do the same job as a flash, with better results.

Very few animals look good from a high angle; if you can take the camera down to their eye level the photo will often be better (admittedly that represents a challenge for iguanas, and beware of a suspicious lunge from a sleepy sea lion!). A bean bag or mini-tripod will help you get the camera down low, and lightweight-foam knee pads allow you to check the viewfinder in comfort. Tripods and walking-stick mounts with a cable release help to keep the camera stable, which improves sharpness. (You may only appreciate this later when you get your pictures printed for the wall or your video projected for friends!)

VISITOR SITES

The following selection of visitor sites are often offered on boat tours. The sailing distances between these sites partly determines which of them can be combined on a single itinerary.

Punta Suarez, Española

At between 3 and 5 million years old, Española is considered the oldest island in Galápagos. Having drifted southeast of the new active volcanoes, it is now a remnant of its former self. But what it lacks in height and volcanic activity it makes up for in spectacular and abundant wildlife. Sea lions may surf the translucent waves in front of the jetty that marks the start of the trail. The most inquisitive bird in Galápagos, the hood mockingbird, will undoubtedly attract your immediate attention, and all around marine iguanas will be soaking up the sun.

The first section of trail comprises numerous white beaches full of sea lions. At any point along the way you will notice the largest lava lizards in the islands. Galápagos doves and three species of finch go about their daily routines even when a Galápagos hawk is perched near by. The looped trail passes along a section of seaside cliffs and over open platforms of weather-rounded rock. Halfway through is a section of coast exposed to heavy swells where the famous blow hole can jettison a spray 30 m into the air.

In early April the first subtle effects of the cool season are felt as waved albatross arrive *en masse* to start breeding. Española also has one of the largest colonies of blue-footed boobies, and numerous Nazca boobies, swallow-tailed gulls and tropicbirds.

Punta Espinosa, Fernandina

Fernandina's ecology is so pristine that the park allows access to only one visitor site. A shady mangrove forest occupies the first section of the trail, which leads into an open area of brown beach and lava rock, and then splits in two.

One path leads inland over pahoehoe lava to a field of impressive aa. The other follows the coast along beaches and tidal pools, where some of the biggest marine iguanas are found in large numbers. The much smaller lava lizard hangs around and perches on the iguanas' heads to ambush flies in mid-air. When it's hot enough marine iguanas take to the ocean. Dipping into these frigid waters to see them feeding is a rewarding experience, but if you prefer not to brave the elements you can still watch them swim out to sea and disappear under water. Keeping your gaze seaward, you may well see flightless cormorants on fishing forays. All along the coast you can find them hauled out or in nesting mode. If you time it right you will have a ringside seat to watch their snake-necking, body-bobbing mating ritual.

Mariela Islets, Isabela

Lying a few hundred metres from the coast of Isabela are two small islands, favourite nesting grounds for the Galápagos penguin. Mainly in the garúa season patient observation may reveal adults going into their rocky holed nests. Some of these homes are lava tubes. If you are lucky you may be rewarded by seeing a grey downy penguin chick emerge.

Elizabeth Bay, Isabela

Five minutes' boat ride from the Mariela Islets are the enchanting mangrove inlets of Elizabeth Bay, a spectacular example of how efficient and prolific mangroves are at colonizing new lava. You ride through a narrow entrance of lava where tidal waters can rip between the bottle-neck like a river. The shallow water in this entrance is often gin-clear and an excellent window for viewing marine life. Through the gap is a large lagoon of calm water surrounded by mangrove. Green turtles are visible in large numbers: November to February is the best time to see them mating. If the water is clear, rays, sharks and reef fish are often common. The mangrove trees make nesting habitat for pelicans, frigatebirds, herons, finches and warblers, to name a few.

Darwin Bay, Genovesa

This small island is a prime site for seabirds. A wet landing at Darwin Bay reveals white-sand beaches, salt-bush and mangrove habitat, interspersed with intertidal pools. Swallow-tailed gulls adorn the ground. Frigatebirds nest in large colonies on the low-lying salt bush. Males inflate their pouches and shake their wings to attract females to their nests. Red-footed boobies nest mainly in the taller mangroves. The path winds through deep, narrow, intertidal pools where you can often see marine iguanas eating algae, then verges into a short cliff walk where a great view over the crater bay is often accompanied by frigates chasing other birds to steal their food.

Marine iguanas here are much smaller than on other islands, because they receive less algae than other more oceanically productive islands. The pools are some of the best places in the islands to 'armchair snorkel' with iguanas. Looking down into the clear water allows you a rare dry view of them feeding under water. Genovesa also has a soft-spined opuntia cactus that hints at an ecological oddity for Galápagos: no cactus-eating herbivores exist on this island. In fact no land reptiles at all.

There are land birds, however. The Galápagos dove has taken advantage of the cactus's soft spines to eat the pollen and in turn pollinates the flowers. You can also begin to get a handle on identifying the finches, as there are only four species here.

Prince Philip's Steps, Genovesa

A steep walk up lava steps takes you on to a cliff, where you pass Nazca and red-footed boobies before entering a palo santo forest rich with its pungent sandalwood aroma. The tree roots are a spectacular example of mechanical erosion of lava and make a beautiful organic contrast to the brittle lava they break down. Red-footed boobies nest here and are a constant companion until you emerge into open ground, where a trail follows the coast and takes you past an oxidized brown-lava platform leading out over a 40-m cliff. The challenge is to spot a short-eared owl. Being camouflaged like the brown lava they live by, these birds are hard to spot. It doesn't help that they normally keep very still – except when hunting the storm petrels that fly above the cliffs in vast numbers.

Punta Cormorant, Floreana

The excursion starts with a wet landing on an olivine beach. Olivine is a green, translucent, volcanic crystal that you can find if you look carefully through the grains of sand. The path splits at the beach-trail head. The right trail follows a beach until you turn inland and walk over flat cinder gravel, where you'll notice an unusual bushy plant called *Lecocarpus pinnatifidus* (found only on Floreana). Approximately 100 m inland the path arrives at a lagoon where greater flamingos can be seen wading through shallow brackish water to feed. In the warm season you may be lucky enough to see their spectacular courtship dance.

The other path overlooks the lagoon and follows through inland, arid zone-vegetation. Eventually it arrives at a beautiful white sand beach surrounded by a shallow turquoise bay – a great place to see turtles and rays in the shallows.

Post Office Bay, Floreana

Here you will be retracing Darwin's footsteps. A brief walk leads you inland to the famous post-office barrel and the historic remains of previous occupants involved in the gruesome whale-oil extracting business. The large, rusty boiling vats lie abandoned. The path continues on to lava tubes and a view of the island from an eroded cinder cone.

North Seymour

A dry landing puts you on to red-brown rocks scattered with sea lions and swallow-tailed gulls, then a circular path leads over a boulder ground to a beach. Here marine iguanas, blue-footed boobies and sea lions are prime candidates for your attention, unless you hit the surfing season between January and April. Then you will be mesmerized by sea lions rocketing through the translucent waves. If you can pull yourself away, you will see both magnificent and great frigatebird colonies close up. A circular walk back to where you started may reward you with a sighting of the biggest land iguanas in Galápagos.

Rábida

The rust-red beach is an unusual feature in the Galápagos landscape. Sea lions are found here in large numbers. A lagoon behind the beach attracts sea lions and may reveal flamingos and other lagoon birds such as stilts and white-cheeked pintail ducks. A walk through arid vegetation provides a picturesque view of the beach. Finches, Galápagos doves and mockingbirds are also around.

Bartolomé

The view of Bartolomé surrounded by shallow blue waters, with Pinnacle Rock as the centrepiece to the dramatic volcanic backdrop of Santiago, is a must-see. The most popular option is to hike to the top; on the way you will see many well-preserved volcanic features, including parasitic cones, lava tubes and spatter cones.

Alternatively, you can walk from the north beach to the south, traversing a small sand dune on the way. The south beach often has white-tip reef sharks and rays lying in the surf-zone shallows. You may also see penguins flying around under water, foraging as they go.

Sullivan Bay, Santiago

A walk on a relatively young lava flow provides a fascinating look at how lava cools into extraordinary designs and allows pioneer plants to establish on its surface. The trail loops around a large iron-oxidized cinder cone, which makes an impressive contrast between the old and the new jet-black lava. This surreal landscape is the next-best thing to walking on the moon.

Puerto Egas, Santiago

The most popular walk here takes you to some ash cliffs that have been eroded into fascinating shapes. The ash and lava rock have also forged impressive pools that harbour many intertidal invertebrates and fish. Marine iguanas, sea lions and wading birds such as oystercatchers, whimbrels and various heron species are common. The main feature is the series of lovely grottoes – collapsed lava tubes that provide a perfect home for the smallest pinniped in the world, the Galápagos fur seal.

Gardner Bay, Española

With its surrounding turquoise waters this lovely beach looks like somewhere in the Bahamas. A wet landing will be followed by curious-as-ever hood mockingbirds coming up to you and investigating what you have to offer. Of course park rules will leave them going away empty 'handed'.

Sea lions live in partitioned colonies along the beach. In the breeding season males actively defend their territories, constantly chasing and fighting one another. Turtles are often seen in shallow water. In the breeding season you can sometimes see them mating in the surf zone. Galápagos hawks may stop in for a look. This is a good opportunity to explore on your own, as the guides permit you to walk alone on this beach.

Tagus Cove, Isabela

An ash trail leads around Darwin Lake, which is in a tuff cone on the edge of a palo santo forest. Darwin's finches are numerous. You climb to a stunning view overlooking Darwin volcano, where, in addition, you gain an insight into what it feels like to be standing on the flanks of a great volcano, looking out over the landscape that its eruptions have sculpted over the years.

Los Gemelos, Santa Cruz

Easily accessed by the main road, this impressive pair of pit craters is surrounded by an endangered endemic forest of *Scalesia pedunculata* trees. A walk through the forest adjacent to the larger crater may reveal the famous woodpecker finch and other tree finches. Not many people get to see a woodpecker finch using a tool to extract insects from holes, but this is a place where patience may reward you. You don't need so much luck to see the lovely vermilion flycatcher, whose crimson feathers advertise its presence very conspicuously. The more cryptic short-eared owl may surprise you if you keep a sharp look-out: it normally perches on the trees looking for mice and finches to eat. The walk ends with a vista overlooking the pit crater and canopy: a wonderful view that on a clear day reveals many of the other islands.

Darwin Station, Santa Cruz

Your chance to see conservation in action. The captive tortoise-rearing centre has been very successful breeding and reintroducing tortoises back to the islands. A boardwalk passes hatchlings in their pens and the famous Lonesome George (see page 170). The trail continues to other tortoises from various islands, including the extremely saddle-backed subspecies from Española. In one pen you can mingle with a group of large dome-shaped tortoises as they go about their daily routine.

El Junco, San Cristóbal

A drive from the main port of Puerto Baquerizo Moreno gets you to one of the very few fresh-water lakes in Galápagos. Then a ten-minute walk up a well-built trail rewards you with a unique view of this unusual feature. Your visit may coincide with frigatebirds washing themselves in the lake. You may also encounter common gallinules, white-cheeked pintail ducks and stilts, among others. On a good day you can see all the way down to the coast.

El Chato, Santa Cruz

It is a 25-minute drive from Puerto Ayora to the car park above the tortoise reserve. From there you walk 20 minutes through humid-zone vegetation looking for vermilion and Galápagos flycatchers. This band of vegetation is more typical of higher elevations, where moisture from clouds drips over the landscape and incites lush growth. Hidden in the foliage you may encounter a giant tortoise, but if not the chances are you will see them when you reach the fresh-water ponds further down the trail. White-cheeked pintail ducks, gallinules and both rail species can also be seen here and frigatebirds sometimes turn up to dip in the pools.

South Plaza

It is a marvel how much wildlife exists on this tiny (1-km long, 130-m wide), uplifted island. Immediately after landing on a concrete dock you are surrounded by a bustling colony of sea lions. Just off the jetty marine and land iguanas can be seen in the same area. A circular trail leads to a cliff, passing opuntia cactus and *Sesuvium*, a ground-matted succulent that turns a spectacular red in the dry season. In the wet season the beautiful flowers of another succulent, portulaca, match the yellow land iguanas. On the cliff walk you will see swallow-tailed gulls, Galápagos shearwaters, blue-footed boobies, Nazca boobies, frigatebirds and red-billed tropicbirds.

Shearwaters are particularly impressive to watch in their agile aerobatics. They fly in and out of the cliffs at breakneck speeds. At the end of the island you enter a bachelor colony of sea lions resting on slabs of basalt that have been rubbed smooth by generations of wear. It is amazing to watch sea lions negotiate the steep cliff they have to climb up to reach their resting places.

Santa Fe

A very picturesque anchorage with shallow turquoise water and sheltered white beaches that are a favourite haul-out for sea lions. The trail leads from the beach over lava boulders across a small cliff where the largest-diameter opuntia cactuses stand impressively tall and wide. Normally you can find the pale land iguanas that are endemic to this island: if a cactus pad or fruit drops to the ground they will be on to it in a flash.

The trail loops through more land-iguana territory surrounded by arid vegetation. At any point you may be lucky enough to see an endemic rice rat whose relatives have been exterminated on most other islands: normally nocturnal, they can occasionally be seen by day. The walk ends at the other beach where you may have to walk carefully around a dense sea-lion colony.

DIVE SITES

The following is a selection of the sites that make diving in Galápagos such a unique experience.

Darwin's Arch, off Darwin

Galápagos's most remote dive site; along with its close neighbour, Wolf, considered the *crème de la crème* of Galápagos diving. From the onset you will be stunned by the profusion of life that graces the reefs and surrounding ocean of this isolated oasis. The first section is made up of mini-walls, ledges and platforms that border the ocean until the volcanic rock merges with shallow coral reef, one of the largest unbroken strands to exist in the islands, and home to many colourful tropical reef fish. Darwin also provides a prolific food source for larger creatures: walls of hammerheads cruise through the water column and over the reef. Sightings of whale sharks are common from June to November.

Wolf

A world-class dive site, the most reliable place to witness the epic sight of large schools of scalloped hammerheads and Galápagos sharks, often in their hundreds. The boulder field on the eastern side is the best site for hammerhead schools and the cleaning stations of butterfly fish that gather to pick parasites from the cruising sharks. Free-swimming moray eels can occasionally be seen hunting here during the day.

An exciting alternative site is the 'pinaculo' on the northeast tip of the island, where you can swim up through a chimney to be spat out into the strong current on the pinnacle of rock. It is a convoluted route, only to be attempted by experienced divers under good supervision.

When planning a trip to either of these islands, book well in advance and don't expect to be alone. Wolf and Darwin simply don't disappoint!

Roca Redonda

For experienced divers only, this isolated outpost of dramatic vertical rock is home to a 'beehive' of seabirds – the first hint of the profusion of marine life below. On the southeastern side, small pockets of gaseous bubbles indicate that the site is on the tip of a monumental volcano, plunging to abyssal depths. A boulder slope of about 40-m width is home to a mix of tropical and cold-water species. The rocky bottom is covered with a dense matting of barnacles that filter feed on the prolific phytoplankton. Galápagos reef sharks school near the island and hammerheads cruise the water column just beyond the rocky slope. Vast schools of barracuda and steel pampano are often seen being chased by sea lions, and fur seals hang out near the surge zone.

Punta Vicente Roca – The Ice Box!

Nestled in the northwest crook of Isabela, under monumental cliffs that are remnants of a volcano that split in half and fell into the ocean. The dramatic vertical topography extends under water. A well-protected cove allows divers to spend time close to the wall where a profusion of invertebrate life is concentrated. The most conspicuous inhabitants are the gorgonians, whose geometric stems appear more plant-like than animal. Corals and anemones are also found in large numbers. The more cryptic but rewarding inhabitants include the camouflaging frogfish. Open-ocean life may also boggle your goggles as turtles, rays, sea lions and mola mola glide by in the silence.

Tagus Cove, Isabela: The Anchorage

A perfect dive for beginners and experts alike, the sandy bottom here is host to a surprisingly large number of interesting creatures, notably the red-lipped batfish with its strange leg-like fins, unicorn-shaped head and red lips. Anemones flutter in the gentle roll of the ocean and make a picturesque pattern of an otherwise monotonous brown carpet of sand. But look closely and surprises will appear, the most spectacular being the breeding male pipe blenny who emerges from a tubeworm and displays his exquisitely coloured dorsal coat to attract a female. Numerous other species use the sand as a hiding place to avoid predation or conversely to prey on unsuspecting passers-by. The rainbow tonguefish buries itself in the sand at the slightest provocation but can be seen lying in wait to pounce on prey. The delightful seahorse is also present.

Tagus Cove: The Outer Wall

The outer lip of the protected cove is a very rewarding dive for those willing to make slightly more effort to get there. An enchanting garden of endemic black coral is attached to boulders lying among carpets of smooth sand. Hidden among the ledges of the walls are many different types of invertebrates, which in turn attract the small horn shark. Turtles are often found sleeping on the ledges and will even ignore divers who don't approach too closely. The sandy substrate attracts bivalves and the predators that feed on them. It's common to see stingrays, eagle rays and others nosing around the sand for a meal.

Cape Marshall, off eastern Isabela

The best place for shark sightings in the central islands. The underwater topography is dominated by young lava flows that cascade abruptly to great depths, forming geometric blocks of hexagonal pillars – some of the best examples found anywhere under water. Along the shallow margins, the current sweeps over the volcanic landscape and allows divers to drift effortlessly. Huge schools of fish inhabit the protected reef. Of particular interest is the endemic brown-striped salema that shoals in spectacular aggregations that can momentarily block out the sun. Sea lions sometimes chase through the schools of salema, creating chaos. Manta rays, however, are the most common spectacle and often plough the reef for plankton or linger to be cleaned. In the open ocean, any moment can be rewarded with jaw-dropping sightings of schooling hammerheads, Galápagos sharks or even orca pods that patrol the coast in search of quarry.

Seymour

At the eastern point of the island the sloping boulder fields and gullies lead around to rock platforms and underwater cliffs of basalt columns. Snapper, jacks and the occasional massive school of salemas gather here, but the highlight is the schooling white-tip reef sharks that hang in the current and allows divers within arm's length. During February and March keep an eye open for females with distended bellies soon to pup. You may also see hammerheads and Galápagos sharks if you keep an eye off into the blue. Further south, in the channel between Mosquera and Seymour, the seabed turns to white sand, offering sightings of large swathes of garden eels and eagle, marbled or cow-nosed rays feeding. The beautifully patterned tiger snake eel also commonly hunts across these sands.

Cousins Rock

A gem of a site, accessible by day-dive boats from Santa Cruz. Above the surface a diagonal stratum of rock provides excellent sleeping ledges for sea lions and fur seals. An underwater ridge drops away to deep water and, as seen above the surface, is made up of shelves creating the perfect shelter for all manner of critters. The walls around the north side of the islands offer yet another habitat. Seahorses can be seen in black coral, as can the endemic blue-striped sea nudibranch. On the west side of the underwater ridge the slope is steadier and sandy and large schools of Creole fish, pelagics and Pacific barracuda can be seen. Under the pointed tip of the island sea lions prey on these schools – a spectacular show for those privileged to witness it.

Gordon Rocks

Reputed to be the best dive in the central islands, Gordon is accessible by day boats from Puerto Ayora and dedicated dive charters. Two prominent rocks form the semicircle of an eroded tuff cone off the coast of the Plazas. Strong currents make this an unsuitable dive for beginners but a Mecca for marine life. Eagle rays, stingrays, reef fish, pelagic fish, turtles, hammerheads and Galápagos sharks all gather here. But Galápagos diving is not all about sharks! Look around the barnacle-encrusted rocks for the exquisite endemic barnacle blenny, which makes its home in disused barnacle shells, darting out into the water column to catch morsels of plankton that drift past.

Champion, Floreana

The dive begins on the wall section of the island where black corals, schools of pelagic fish and rays make for a classic Galápagos underwater spectacle. While drifting along with the current, keep your eye open for feeding sea lions cruising the wall. As you round the southerly point the seabed becomes shallower and you reach the sea-lion colony. The boulder beach serves as a nursery for pups and it is usual to be joined by inquisitive, playful youngsters. The ballet they perform in front of you will put any of your acrobatic efforts in heavy scuba to shame!

Devil's Crown, off Floreana

An almost perfectly circular eroded crater, surrounded by shallow turquoise seas. The dive normally begins over sand flats where a unique endemic pebble coral is found strewn over the sand. A typically swift current sweeps divers on to the outside of the crater, where most of the fish life is concentrated. Neat pockets of reef fish gather in tight aggregations. The rocks are covered with less conspicuous but no less abundant life. Rainbow wrasse, barnacle blennies, sabretooth wrasse and many others eke out a living among the remains of the crater walls that now lie scattered over a white sandy seabed. Rays, white-tip reef sharks, turtles and even small schools or individual hammerheads glide by.

Enderby Islet, off Floreana

The young volcanic cone has steep crescent-shaped walls that plunge into the abyssal depths. The dive normally begins inside the crater and follows a thin spindle of ridge out to open ocean. The current hits the ridge and provides an oxygen-rich environment for a myriad of reef fish. In various sections along the way, turtles seek out fish to clean parasites. Galápagos reef sharks patrol inside the ridge and hammerheads cruise the open ocean just outside the crater. Following the outside of the crater, the ridge turns into a steep invertebrate-encrusted wall, where closer inspection reveals a pock-marked surface that gives protection to a host of fish and invertebrate life.

El Bajo, Gardner Bay, Española

This site has a great diversity of fish life and with the shallow, sandy seabed reflecting a lot of light, you'd be forgiven for thinking you'd drifted into colourful Caribbean waters. El Bajo is a rocky table, standing up 3–4 m from the sandy seabed, and at a leisurely swim can be circumnavigated during the course of a dive. Given the good light, consistently shallow depths and often clear visibility, it is a good site for underwater photography. Schools of reef fish and rays are attracted to El Bajo to feed from the sand about its base. White-tip reef sharks can regularly be seen resting under ledges. Groups of butterfly fish indicate cleaning stations that are worth watching and on the table top of rock green turtles can be seen hanging in the water column, being picked free of dead skin and parasites.

WHERE TO STAY

An alternative to the live-aboard boat trip is to base yourself in one of Galápagos's few towns and villages, which offer hotels or guest houses and the opportunity to book day trips to see the wildlife on land and sea.

Puerto Villamil, Isabela

Estimated population: 3500

Situated on the southern flanks of Sierra Negra volcano, nestled among palm groves and spread along a beautiful white-sand beach and turquoise sea, this is perhaps the most charming of all Galápagos settlements. Sand streets and a simple low-key atmosphere make it a relaxing and pleasant destination. Primarily a fishing village known locally as Tierra de Nadie (land of nobody), Puerto Villamil remained a backwater until very recently. Now one up-scale hotel and several guesthouses cater to a growing tourist presence.

There are also some interesting sites to visit near by. A 45-minute drive takes you to the base of Sierra Negra, one of the most active volcanoes in the world; another hour's horse ride gets you to the rim. By boat you can visit mangroves, offshore islands, beaches and a fascinating lava-tunnel labyrinth full of marine life. If you want closer access to wildlife near town, there are some excellent choices, ranging from flamingo lagoons and sea-lion and penguin mangrove pools to offshore islets where you can find marine iguanas, white-tip reef sharks and a host of other wildlife.

Puerto Ayora, Santa Cruz

Estimated population: 15,000

The commercial capital and largest town in Galápagos, this is where most tour operators base their businesses. The majority of tour boats stop here to see the Charles Darwin Station and National Park captive tortoise-rearing programme. The main streets are lined with shops, restaurants and bars catering to tourists and locals. This is also an excellent place to book day tours to visit nearby islands or go diving, and there is no shortage of white pick-up taxis to take you around the island. Hotels range in price from five-star luxury to the most basic.

Puerto Baquerizo Moreno, San Cristóbal

Estimated population: 10,000

The governmental capital of the province, this pleasant seaside settlement used to be the economic hub of Galápagos when coffee, cattle and sugar were exported to mainland Ecuador. Today it is an interesting if awkward hybrid of tourist development and fishing village. There is an informative visitor centre and several trails leading out to arid habitat and beach. Surfers from all over the world share the waves with surfing sea lions. Several hotels now offer reasonable accommodation, while the few restaurants and bars cater to low-key tourism with local flair.

Puerto Velasco Ibarra, Floreana

Population: 120

Anyone wishing to witness an authentic down-to-earth side of Galápagos life should consider visiting this pleasant backwater. There is one beach, one hotel, one restaurant and one bar doubling as the shop. There is not much to do, but it is a stop-off point for the local history – Charles Darwin visited and numerous Galápagos characters have lived here over the years. The Wittmer family, who lived in a cave when they first arrived in the 1930s, now own the hotel.

HOW YOU CAN HELP

Some 30,000 people already live on five islands of Galápagos and the population is rising at about 5.7 per cent a year. This rise can mainly be attributed to the rapid growth in tourism – over 120,000 people visit each year and the numbers are increasing. Our money provides a motive and an economic means for people to live in Galápagos. They need fuel, electricity, food and materials in order to sustain themselves and the vast majority of these goods are shipped to the islands from the mainland. A direct result of this need is the introduction of alien species that severely threaten the native flora and fauna. Every one of us who visits this fragile ecosystem may be contributing – at least indirectly – to its damage. So if we can find ways to offset our risk as individuals by contributing more to the conservation of the islands, Galápagos will be better for it.

Non-monetary contributions

◆ Ensure you don't bring any alien species to Galápagos.
◆ Follow established rules while you are in the park.
◆ Take pictures or videos of any activity you feel is compromising the integrity of the ecosystem within the park and send them to the Director of the National Park and the Communications Director of the National Park (address below). Photos are necessary as evidence if the park is to pursue any legal action.
◆ There is a blanket ban on fishing by tour boats everywhere in the islands, with no exceptions. If you witness any illegal fishing, again please photograph it and report it to the Director of the National Park.
◆ Visit the local non-governmental organizations and find out how you can contribute to their work in Galápagos either as a volunteer, as a professional or in other ways.

Financial contributions

Profits from all items sold in the park and Darwin Station shops go to conservation. The following is a list of organizations you can support with monetary donations:

Charles Darwin Research Station (CDRS):
www.darwinfoundation.org/
Galápagos National Park:
www.galapagospark.org
Galápagos Municipal Library
WildAid: www.wildaid.org
WWF: www.worldwildlife.org/galapagos
Sea Shepherd Conservation Society:
www.seashepherd.org
Galápagos Conservation Trust (GCT):
www.gct.org.uk
Galápagos Conservancy: www.galapagos.org

You can contact the Director of the National Park by writing or emailing (but please no general tourist enquiries):

Director de Parque Nacional de Galápagos
Puerto Ayora
Santa Cruz
Galápagos
Ecuador

Email: info@spng.org.ec
For complaints or reports of illegal activity, email denuncias@spng.org.ec

FURTHER READING

Naturalist Guidebooks

Constant, P., *Galápagos: A Natural History Guide* (Odyssey, 1995).
A visitor favourite, nicely written and covering a good range of species.

Constant, P., *Marine Life of the Galápagos* (Sing Cheong Printing, 1992).

Fitter, J., Fitter, D., and Hosking, D., *Wildlife of the Galápagos* (Collins Safari Guides, 2000).
An excellent photographic field guide to the most commonly encountered species, it fits easily into a day sack. Don't leave the boat without it!

Horwell, D., and Oxford, P., *Galápagos Wildlife* (Bradt, UK, and Globe Pequot, USA, 1999).
Guidebook with good detail about species and visitor sites. Also look for P. Oxford and R. Bishe's mesmerizing photographic book on Ecuador.

Jackson, M.H., *Galápagos: A Natural History* (Calgary, 1993).
A long-standing favourite for its reliable, no-nonsense information.

McMullen, C.K., *Flowering Plants of Galápagos* (Cornell, 1999).
A specialist but accessible text, allowing the identification of many Galápagos plants.
A must for plant lovers.

Swash, A., and Still, R., *Birds, Mammals & Reptiles of the Galápagos Islands* (Yale, 2000, and published in the UK by Pica Press and Wild Guide).

Photo Essays and Souvenirs

Anhalzer, J.J., *Galápagos* (Anhalzer, 1998).
Watch out for this and other Anhalzer photographic books while in Ecuador. The aerial photos in particular are unrivalled.

De Roy, T., *Islands Born of Fire* (Warwick House, 2000).
Remarkable photos and a moving text from Tui De Roy's lifelong experience of the islands. A beautiful book.

De Roy, T., *Galápagos Islands Lost in Time* (Viking, 1980).
A photo collection representing the first ten years of her work, with a candid account of exploration and discovery.

Popular Science

Darwin, C., *Voyage of the Beagle* (Henry Colburn, 1839. Various editions available, including Penguin Books, 1989).
The fascinating first-hand account of Darwin's world voyage includes a chapter on the Galápagos – and a few hints that Darwin was already starting to wonder about Galápagos evolution 20 years before the publication of *Origin of Species*.

Darwin, C., *On the Origin of Species – by Means of Natural Selection* (John Murray, 1859. Various editions available, including Penguin Books 1985).
Still remarkably readable and generally accurate. Confirmation, if it were needed, of the author's methodical genius. Read it on Galápagos (but avoid the temptation to quote him at dinner). Don't be disappointed that Galápagos is not the basis of his account. The islands were the origin, but ultimately not the proof, of all his views.

Krischer, J.C., *Galápagos* (Smithsonian, 2002).
A very readable series of essays on natural-history topics centred around these islands.

Larson, E.J., *Evolution's Workshop: God and Science on the Galápagos Islands* (Basic Books, US, 2001, and Penguin, UK, 2002).
A wonderfully written and detailed history of the islands, putting them in context and revealing their greater impact on the world beyond. Recommended.

Perry, R. (ed.), *Key Environments: Galápagos* (Pergamon, 1984).
Very useful collection of individual essays by eminent scientists about Galápagos geology, climate, flora and fauna. Sadly hard to find.

Weiner, J., *The Beak of the Finch: A Story of Evolution in Our Time* (Vintage Reprint edition, 1995).
Pulitzer Prize-winning book describing how one group of Galápagos animals, the finches, allows us to understand mechanisms in evolution.

Human History

D'Orso, M., *Plundering Paradise: the Hand of Man on the Galápagos Islands* (Perennial, 2003).
Interesting journalistic treatment of the issues and characters faced by the islands at the start of the new millennium. Read this if you want to glimpse the darker side of the islands' social fabric.

Hickman, J., *The Enchanted Islands: The Galápagos Discovered* (Anthony Nelson, 1985).
A very enjoyable account of the earlier human history of these islands.

Idrovo, H., *Footsteps in Paradise* (Ediciones Libri Mundi/Enrique Grosse-Luemen, 2005).
A well-written and finely illustrated book covering the recent Spanish-speaking history of the islands in detail. Helps to redress the rather Euro-centric publishing record of the islands' history.

Latorre, O., *The Curse of the Giant Tortoise: tragedies, crimes and mysteries in the Galápagos Islands* (National Cultural Fund, 1997).
The result of a lifetime's dedicated interest in the human history of Galápagos, this locally available book, also available in the USA, is on the list for anyone wanting to get deeper in to the subject.

Treherne, J., *The Galápagos Affair* (Pimlico, 2002).
The deaths on Floreana in the 1930s read more like an Agatha Christie novel than history, but this account gives an insight into how people can make hell for one another while looking for paradise.

Wittmer, M., *Floreana* (Anthony Nelson, 1989).
This book gives an insight into some remarkable characters and times on the islands, the like of which were never to be seen again. Margret Wittmer was as elusive in print as she was in life – never giving a clear opinion on her island's whodunit.

Woram, J., Charles *Darwin Slept Here: Tales of Human History at World's End* (Rockville, 2005).
An account of the history of the islands – focusing on the many historical inaccuracies and suppositions to be found in earlier accounts and local lore.

Just for Fun

Angermeyer, J., *My Father's Island: A Galápagos Quest* (Pelican, 2003).
A book of local interest that also manages to be a fine piece of literature in its own right. It gives a great insight into life of some of the early European settlers on Santa Cruz.

Melville, H., *The Encantadas and Other Stories* (Dover, 1854).
Melville's homage to his love–hate relationship with these islands. His further experiences led him to write his classic, *Moby Dick*.

Vonnegut, K., *Galápagos* (Delacorte, USA, and Cape, UK, 1985).
Not always rated as his greatest book by die-hard Vonnegut fans, but entertaining reading nonetheless. Big brains beware!
Mere opinions, in fact, were as likely to govern people's actions as hard evidence, and were subject to sudden reversals as hard evidence could never be. So the Galápagos Islands could be hell in one moment and heaven in the next, and ... Ecuadorian paper money could be traded for food, shelter, and clothing in one moment and line the bottom of a birdcage in the next, and the universe could be created by God Almighty in one moment and by a big explosion in the next--and on and on.

BIBLIOGRAPHY

In addition to the more popular titles listed above, anyone wanting to take their Galápagos studies further is recommended to seek out the following books and papers.

CHAPTER 1

Caccone, A., et al., 'Origin and evolutionary relationships of giant Galápagos tortoises', *PNAS* (9 Nov, 1999), vol. 96, no. 23, pp. 13223–8.

Geist, D., 'On the emergence and submergence of the Galápagos islands, Ecuador', *Noticias de Galápagos* (1996), vol. 56, p. 5.

Harp, K., and Geist, D., 'Galápagos Plumology', *Noticias de Galápagos* (1998), vol. 59, p. 23.

Hoernle, K., et al, 'Missing history (16–71 Ma) of the Galápagos hotspot: implications for the tectonic and biological evolution of the Americas', *Geology* (September 2002), vol. 30, no. 9, pp. 795–8.

Lamb, S., and Sington, D., *Earth Story: The Shaping of our World* (BBC Books, 2003).

Simkin, T., and Siebert, L., *Volcanoes of the World* (2nd edition: Geoscience Press, 1994).

Sallares, V., et al., 'Seismic structure of the Carnegie ridge and the nature of the Galápagos hotspot', *Geophys. J. Int.* (2005), vol. 161, pp. 763–88.

Tye, A., and Aldaz., 'Effects of the 1997–1998 El Niño event on the vegetation of Galápagos', *Noticias de Galápagos* (1999), vol. 60, pp. 22–4.

CHAPTER 2

Beebe, W., *Galápagos: World's End* (Putnam, 1924. Modern reprint Dover Publications, 1988).

Hoff, S., *Drommen om Galápagos: En Ukjent Norsk Ultvandrerhistorie* ('The Galápagos dream: an unknown history of Norwegian emigration') (Grodahl, Oslo, 1985).

Philbrick, N., *In the Heart of the Sea: The Tragedy of the Whaleship Essex* (Penguin Putnam, 2000).

Porter, D., et al., *Journal of a Cruise Made to the Pacific Ocean* (Classics of Naval Literature, Naval Institute Press, 1986).

Preston, D., and Preston M., *A Pirate of Exquisite Mind: Explorer, Naturalist and Buccaneer: The Life of William Dampier* (Walker, 2004).

CHAPTER 3

Browne, E.J., *Charles Darwin: Voyaging* (Princeton, 1996).

Darwin., C., *The Autobiography of Charles Darwin 1809–1882* (Norton, 1993).

Dawkins, R., *The Blind Watchmaker* (Norton, 1986).

Dennett, D.C., *Darwin's Dangerous Idea: Evolution and the Meanings of Life* (Simon & Schuster, 1996).

Estes, G., Grant, T., and Grant, P., 'Darwin in Galápagos: his footsteps through the archipelago', *Notes Rec. R. Soc. Lond.* (2000), vol. 54, no. 3, pp. 343–68.

Grant, P.R., *Ecology and Evolution of Darwin's Finches* (Princeton, 1986).

Lack, D., *Darwin's Finches* (Cambridge, 1947. Various editions available, including Cambridge, 1983).

Lyell, C., *Principles of Geology* (John Murray, 1830. Various editions available, including Penguin Classics, abridged edition, 1998).

Malthus, T.R., *An Essay on the Principle of Population* (1798. Various editions available, including Penguin Classics, 1983).

Mayr, E., *Animal Species and Evolution* (Belknap, 1963).

Paley, W., *Natural Theology* (Kessinger, 2003).

Sulloway, F.J., 'Darwin and his finches: the evolution of a legend', *Journal of the History of Biology* (1982), vol. 15, pp. 1–53.

Sulloway, F.J., 'Darwin's conversion: the *Beagle* voyage and its aftermath', *Journal of the History of Biology* (1982), vol. 15, pp. 325–96.

Wilson, E.O., *Biodiversity* (National Academies Press, 1988).

CHAPTER 4

Chambers, P., *A Sheltered Life: The Unexpected History of the Giant Tortoise* (John Murray, 2004).

Hayashi, A.M., 'Attack of the fire ants', *Scientific American* (1999), vol. 280, pp. 26–8.

MacArthur, H., and Wilson, E.O., *The Theory of Island Biogeography* (Princeton, 2001).

Pritchard, P.C.H., 'The Galápagos Tortoises: nomenclatural and survival status', *Chelonian Research Monographs* (1996), no. 1.

Quammen, D., *Song of the Dodo* (Prentice Hall & IBD, 1996).

CHAPTER 5

Castillo-Briceño, P., Marquez, C., Widenfield, D., Snell, H., and Jaramillo, A., 'El Niño – Southern Oscillation and Health Status of Galápagos Marine Iguana', *Ecologia Aplicada* (2004), vol. 3, no. 1, p. 2.

Castro, I., and Phillips, A., *A Guide to the Birds of the Galápagos Islands* (Princeton, 1996).

Eibl Eibesfeldt, I., 'Part 1: Natural history of the Galápagos sea lion', in *Key Environments: Galápagos*, ed. Perry, R. (Pergamon, 1984).

Wikelski, M., and Hau, M., 'Is there an endogenous tidal foraging rhythm in marine iguanas?', *Journal of Biological Rhythms* (1995), vol. 10, no. 4, pp. 335–50.

Wikelski, M., Carrillo, V., and Trillmich, F., 'Energy limits to body size in a grazing reptile, the Galápagos marine iguana', *Ecology* (1997), vol. 78, no. 7, pp. 2204–17.

Wikelski, M., and Thom, C., 'Marine iguanas shrink to survive El Niño', *Nature* (2000), vol. 403.

CHAPTER 6

Allen, G.R., and Robertson, D. Ross, *Fishes of the Tropical Eastern Pacific* (Hawaii, 1994).

Grove, J., and Levenburg, R.J., *Fishes of the Galápagos Islands* (Stanford, 1997).

Hickman Jr, C.P., *A Field Guide to Sea Stars and Other Echinoderms of Galápagos* (Sugar Spring Press, 1998).

Hickman Jr, C.P., and Finet, Y., *A Field Guide to Marine Molluscs of Galápagos* (Sugar Spring Press, 1999).

Hickman Jr, C.P., and Zimmerman, T.L., *A Field Guide to Crustaceans of Galápagos* (Sugar Spring Press, 2000).

Humann, P., *Reef Fish Identification: Galápagos* (New World Publications, 1993).

McCosker, J.E., and Rosenblatt, R.H., 'The inshore fish fauna of the Galápagos islands' in *Key Environments: Galápagos*, ed. Perry, R. (Pergamon, 1984).

Merlen, G., *Field Guide to the Fishes of the Galápagos* (Wilmot Books, London, 1988 & Libri Mundi, Quito, 1988).

Pauly, D., *Darwin's Fishes: an Encyclopedia of Ichthyology, Ecology and Evolution* (Cambridge, 2004).

Wellington, G.M., 'Marine Environment and Protection' in *Key Environments: Galápagos*, ed. Perry, R. (Pergamon, 1984).

Wellington, G.M., *The Galápagos Coastal Marine Environment*, unpublished manuscript, Charles Darwin Research Station Library, Galápagos, 1975.

CHAPTER 7

Bahn, P., and Flenley, J., *Easter Island Earth Island* (Thames and Hudson, 1992).

Carlquist, S., *Island Biology* (Columbia, 1978).

Diamond, J., *Collapse: How Societies Choose to Fail or Succeed* (Viking, 2004).

Abstracts and full texts of a wealth of excellent articles are available via the Darwin Foundation website, currently at www.darwinfoundation.org/articles/index.html

ACKNOWLEDGEMENTS

This book, and the BBC series *Galápagos* that it accompanies, has been a collaboration of authors, friends and experts. The kindness and generosity shown us while writing it cannot easily be expressed or repaid. The Government of Ecuador, the Ministro del Exterior and the Ecuadorian Embassy in London and Quito have given support, permissions and help throughout this project. We hope their commitment to the wildlife of these islands will continue undiminished whatever the future brings.

The heads and staff of the National Park and Darwin Station have also shown great backing and patience for our project, sharing with us the hope that by revealing Galápagos's many wonders, the world may the better unite to help save it. The National Park fights daily to look after these islands; their wardens have taken bullets, thorns and insults to keep them safe. Yet even they cannot work the miracle of preserving the Galápagos alone. More than anyone this responsibility has fallen to the people who are now at home on these islands. It is a hard business living in such a delicate paradise, and the international community must help in whatever way they can to find them, and Ecuador, a balance and a future in this precious wilderness. It is to the Parks and the Galápagenos that we dedicate this book.

From our first days on these islands we all owed a great debt of thanks to those naturalists who have over the intervening years taught us what we know, and even helped us discover some things that no one knew. Felipe Cruz, Lenin Cruz, Daniel and Julian Fitter, Gil De Roy, Greg Estes, Thalia Grant, Paul and Gabby MacFarling, Mathias Espinosa, Scott Henderson, Jan Castle and others have shared years, indeed lifetimes, of experience of these enchanted islands.

Others have worked hard with us to make our island voyages possible. In Puerto Ayora: Giok Ong, Tina Fitter, Giancarlo Toti and Veronica Clavijo have looked after logistics. In the field: José Masaquisa, Hugo Ruis, Maria Schumacher, Domingo Gonzalez and the crews of the boats *Pirata*, *Despertad* and *Queen Mabel* have all risked life and limb to get us through rough seas, up steep cliffs and into the remote interiors of these islands. We must in this also single out the great help given to us by the pilots and staff of Project Isabela: Felipe Cruz, Carl Campbell, Tom Poulson, Christian Lavoie, Steve Gamble, Frazer Sutherland, Guchi Fernie, Wilson Cabrerra and Alonso Carrion. They have provided advice and support throughout. Expert assistance under water made our marine work possible and safe; thanks is due to the dive company Scuba Iguana, the boat *Golondrina*, and divers Solón Idrivo, Raphael Gallardo, Josemar Yepez, Dover Medina, Robby Pepola, Jimmy Penarerra, Ruben Idriovo and Juan Carlos 'Macarón' Moncayo.

Film-makers Luke Barnett, Barrie Britton, David Jimmenez, Richard Burton, Peter Scoones, Chris Watson, Jim Clare, David Day and John Waters have joined us on trips, sharing experiences from massed sharks in the deep to erupting volcanoes on the peaks. Thanks for being such good company. In Quito, logistical miracle workers Brigitte Frank, Sandra Serrano and Monica Sanchez of Transcord have helped ease the passage of many jetlagged Englishmen to these islands. Thanks in particular to Sandra, who has cheerfully met requests for everything from video cassettes to giant trampolines.

The scientists of the Darwin Station, including Ros Cameron, Noemi d'Ozouville, Charlotte Caulston, Stuart Banks, Patty Zarate, Alan Tye, Alex Hearn and Christine Parent, have been kind enough to listen to our questions on their areas of expertise and put us right. Many of these scientists, together with Thomas Roedl, Dennis Geist, Jennifer Jacquet, Leticia Valverdes, Sarah Huber and the Galápagos naturalists previously mentioned, have helped with advice and reading of chapters. It goes without saying that any remaining errors are ours and not theirs. In the National Park, Washington Tapia, Lorena Sanchez, Sixto Naranjo, Fabian Oviedo, Victor Carrion, Mario Piu, Juan Chavez, Saul Robalino, Milton Quilligana and Simon Vilamar have all given us their wise counsel. Greg Estes, Thalia Grant and Randal Keynes have given much useful comment on the Darwin chapter.

In BBC Bristol, Neil Nightingale, Mike Gunton, Anna Kington, Liz Toogood and Di Williams have supported us throughout. Melinda Barker, Ted Oakes and Yvonne Ellis were a great help in getting this venture off the ground. Also in the UK, the Galápagos Conservation Trust and Down

House have been kind enough to share their experience and lend their support to this project. Many thanks are also due to our publishers, BBC Books. Shirley Patton, Sarah Reece, Caroline Taggart, Laura Barwick and Andrew Barron have worked hard to make this multi-authored book as coherent and attractive as it is.

It is very clichéd to acknowledge the 'giants upon whose shoulders we climb', but we will anyway. From 'old Dampier' to Darwin, Beebe and the Grants, we feel we know them through their works, and though we can't aspire to their originality, we share their awe of this place. Modern authors and photographers, such as Tui De Roy, Peter Oxford, Daniel Fitter and Gorge Anhalzer, have inspired us all to walk further, climb higher and dive deeper in pursuit of the moments they alone have hitherto captured.

And finally, to the families and loved ones who have played such a key role in our endeavour. Their patience (often sorely tested) with our long trips away and their support during our 'mental absence' while writing this book have been remarkable. We hope that one day all our children will have an opportunity to visit an unspoilt Galápagos and be proud of the book we have written.

Paul D. Stewart, Godfrey Merlen,
Patrick Morris, Andrew Murray,
Joe Stevens and Richard Wollocombe,
July 2006

PICTURE CREDITS

BBC Worldwide would like to thank the following for providing photographs and for permission to reproduce copyright material. While every effort has been made to trace and acknowledge copyright holders, we would like to apologize should there have been any errors or omissions.

Page 1 Tui De Roy; 2 Patrick Morris; 3 & 4 Tui De Roy; 5 Eric Cheng; 6 Daniel Fitter; 7 Tui De Roy; 8 Mitsuaki Iwago/ Minden Pictures/FLPA; 10 Peter Scoones/naturepl.com; 11 Luke Barnett; 12 The Natural History Museum, London; 13 Frans Lanting/Minden Pictures/FLPA; 15 Tui De Roy/ Minden Pictures/FLPA; 16 Yann Arthus-Bertrand/CORBIS; 18 Jeff Schmaltz, MODIS Rapid Response Team, NASA/ GSFC; 19 & 20 Tui De Roy/Minden Pictures/FLPA; 23 Peter Scoones; 24–5 Tui De Roy/Minden Pictures/FLPA; 27 Tui De Roy; 28 top left Frans Lanting/Minden Pictures/FLPA; 28 top right & bottom David Hosking/FLPA; 30 Tui De Roy; 31 Chris Mattison/FLPA; 33 D. Parer & E. Parer-Cook/ AUSCA/Minden Pictures/FLPA; 35 Tui De Roy; 37 Tui De Roy/Minden Pictures/FLPA; 38 Yann Arthus-Bertrand/ CORBIS; 40 Michio Hoshino/Minden Pictures/FLPA; 41 Godfrey Merlen; 42 Paul Stewart; 43 Peter Scoones; 44 Hulton Archive/Getty Images; 45 Tui De Roy/Minden Pictures/FLPA; 46 Paul Stewart; 47 Sloane 45/British Library; 48 Paul Stewart; 49 Flip Nicklin/Minden Pictures/FLPA; 51 Patrick Morris; 52 Richard Wollocombe; 53 & 54 Tui De Roy/ Minden Pictures/FLPA; 56 Bettmann/CORBIS; 59 Richard Wollocombe; 60 & 62 Tui De Roy; 63 Richard Wollocombe; 64 Bettmann/CORBIS; 65 & 66–7 Frans Lanting/Minden Pictures/FLPA; 68 Tui De Roy/Minden Pictures/FLPA; 70 Thomas Roedl; 71 The Natural History Museum, London; 73 Tui De Roy; 75 Randal Keynes; 76 Paul Stewart; 79 Tui De Roy; 81 1570/167/British Library; 83, 84 & 86 Tui De Roy/ Minden Pictures/FLPA; 87 Tui De Roy; 89 Mark Jones; 90 & 91 Tui De Roy; 93 Luke Barnett; 96–7 Tui De Roy/Minden Pictures/FLPA; 99 Tui De Roy; 100 Richard Wollocombe; 101, 102 & 103 Tui De Roy; 106 & 107 Daniel Fitter; 108 Konrad Wothe/Minden Pictures/FLPA; 110, 112 & 114 Tui De Roy/Minden Pictures/FLPA; 117 Mark Jones; 118 Tui De Roy; 119 Tui De Roy/Minden Pictures/FLPA; 120 ARCO/ naturepl.com; 122–3 Tui De Roy; 124 Daniel Fitter; 127 David Middleton/NHPA; 128 Frans Lanting/Minden Pictures/FLPA; 130 David Hosking/FLPA; 134 Tui De Roy; 136–7 Patrick Morris; 138 Chris Newbert/Minden Pictures/ FLPA; 140 Daniel Fitter; 141 Sharon Heald/naturepl.com; 142 Georgette Douwma; 143 Stephen Frink/CORBIS; 144 Pete Oxford/naturepl.com; 146 Peter Scoones; 148 Tui De Roy; 150 Birgitte Wilms/Minden Pictures/FLPA; 151 Tui De Roy; 152 Doug Perrine/naturepl.com; 154 Richard Wollocombe; 155 Tui De Roy; 157 Norbert Wu/Minden Pictures/FLPA; 158 Chris Newbert/Minden Pictures/FLPA; 160 Richard Wollocombe; 162 Godfrey Merlen; 163 & 165 Tui De Roy; 166–7 Pete Oxford/naturepl.com; 169 Godfrey Merlen; 170 Daniel Fitter; 171 Pete Oxford/naturepl.com; 173 Godfrey Merlen; 178, 180 & 181 Tui De Roy/Minden Pictures/FLPA; 183 Daniel Fitter; 185 Mark Jones; 186–7 Tui De Roy/ Minden Pictures/FLPA; 188 David Hosking/FLPA; 190 & 191 both Pete Oxford/naturepl.com; 192 top Tui De Roy; 192 bottom Frans Lanting/Minden Pictures/FLPA; 193 & 194 Pete Oxford/naturepl.com; 195 Tony Heald/naturepl.com; 196 Pete Oxford/naturepl.com; 197 Hans Christoph Kappel/ naturepl.com; 198 both & 199 both Pete Oxford/naturepl. com; 201 top Doug Perrine/naturepl.com; 201 bottom Pete Oxford/naturepl.com; 202 top Christophe Courteau/ naturepl.com; 202 bottom & 203 Pete Oxford/naturepl.com; 205 Mike Read/naturepl.com; 206 top John Cancalosi/ naturepl.com; 206 bottom Rolf Nussbaumer/naturepl.com; 208 Pete Oxford/naturepl.com; 209 Georgette Douwma/ naturepl.com; 210 top Jeff Rotman/naturepl.com; 210 bottom Tony Heald/naturepl.com; 211 Pete Oxford/naturepl. com; 213 Todd Pusser/naturepl.com; 214 Doug Perrine/ naturepl.com; 215 Georgette Douwma/naturepl.com; 216 Fred Bavendam/Minden Pictures/FLPA.

INDEX

Page numbers in *italics* refer to illustrations

COPYRIGHT

This book is published to accompany the television series *Galápagos*, first broadcast on BBC2 in 2006.
Executive producer: Mike Gunton
Series producer: Patrick Morris

Published by BBC Books,
BBC Worldwide Limited, Woodlands,
80 Wood Lane, London W12 0TT.

Published in the United States by
Yale University Press.

ISBN: 978-0-300-12230-5

Library of Congress Control Number: 2006929051

A catalogue record for this book is available from the British Library.

Commissioning Editor: Shirley Patton
Project Editor: Sarah Reece
Copy Editor: Caroline Taggart
Designer: Andrew Barron @ Thextension
Picture Researcher: Laura Barwick
Cartographer: Martin Darlison at Encompass Graphics
Illustrator: Jerry Fowler
Production Controller: Arlene Alexander

Set in Minion and Quay Sans
Colour origination by Butler & Tanner Ltd., Frome, England
Printed by Graphicom Srl, Italy

The paper in this book meets the guidelines for permanence and durability of the Committee on Production Guidelines for Book Longevity of the Council on Library Resources.

10 9 8 7 6 5 4 3 2 1

Page 1: Brown pelicans in Tortuga Bay, Santa Cruz.
Page 2: Clouds tumbling into the caldera on Fernandina.
Page 4: Marine iguanas and a Nazca booby rest on the cliffs of Española.
Page 5: A whale shark cruises among coral fish near Wolf Island.